Springer Polar Sciences

Series editor
James Ford, Department of Geography, McGill University, Montreal, Québec, Canada

Springer Polar Sciences

Springer Polar Sciences is an interdisciplinary book series that is dedicated to research on the Arctic and sub-Arctic regions and Antarctic. The series aims to present a broad platform that will include both the sciences and humanities and to facilitate exchange of knowledge between the various polar science communities.

Topics and perspectives will be broad and will include but not be limited to climate change impacts, environmental change, polar ecology, governance, health, economics, indigenous populations, tourism and resource extraction activities.

Books published in the series will have ready appeal to scientists, students and policy makers.

More information about this series at http://www.springer.com/series/15180

Kirsi Latola • Hannele Savela
Editors

The Interconnected Arctic — UArctic Congress 2016

UARCTIC CONGRESS 2016

SEPTEMBER 12ᵀᴴ – 16ᵀᴴ
ST. PETERSBURG. RUSSIA

UArctic

Springer Open

Editors
Kirsi Latola
Thule Institute, University of Oulu
Oulu, Finland

Hannele Savela
Thule Institute, University of Oulu
Oulu, Finland

ISSN 2510-0475   ISSN 2510-0483   (electronic)
Springer Polar Sciences
ISBN 978-3-319-86186-9   ISBN 978-3-319-57532-2   (eBook)
DOI 10.1007/978-3-319-57532-2

This Springer imprint is published by Springer Nature
The registered company is Springer International Publishing AG
The registered company address is: Gewerbestrasse 11, 6330 Cham, Switzerland

Foreword

In September 2016, Saint Petersburg State University hosted the first ever UArctic Congress. It gathered nearly 500 scientists, experts, students, and leaders from the Circumpolar North over 5 days in 50 sessions. In addition to consuming 5000 cups of coffee, the participants focused on the following key issues:

- The vulnerability of Arctic environments;
- The vulnerability of Arctic societies;
- Local and traditional knowledge;
- Building long-term human capacity;
- New markets the Arctic, including trade, tourism, and transportation.

In addition, the congress included the UArctic Rectors' Forum, the Council of UArctic meeting, the UArctic Student Forum, the UArctic Board of Governors meeting, as well as other organizational events. The congress documented the value of bringing together leading circumpolar scientists from the UArctic Thematic Networks and beyond, with UArctic's leadership and members. This book brings together selected papers from the Science section, the centerpiece of the UArctic Congress 2016.

The University of the Arctic

The UArctic, through cooperation within its very powerful network of over 170 organizations, has demonstrated its ability to deliver multidisciplinary education and research over the past 16 years. This bears great promise for the future of the Arctic and the world. Collaboration through the UArctic helps our members create science-based knowledge on issues crucial for Arctic development which is beneficial for Arctic inhabitants, relevant for the rest of the world, and environmentally, socioculturally, and economically sustainable. The UArctic, through its Thematic Networks and Institutes, can advise decision-makers in policy-making and industry in Arctic economic development to achieve the "triple bottom line" of sustainability:

- The UArctic provides the infrastructure for international science, technology, engineering, art, and mathematics education cooperation, citizen empowerment, and capacity building relevant for the Arctic and the world.
- The UArctic Thematic Networks and Institutes provide a strong basis for shared curriculum and science-based education initiatives, as well as issue-based research collaboration prepared to act on present and emerging needs in and about the Arctic.
- The UArctic represents an operationally unique multilateral infrastructure to enable student and faculty mobility and internships both within and to the Arctic that can be further expanded to meet present and emerging needs.
- The UArctic ensures the best use of present-day and future investments in higher education and research institutions through international collaboration that benefit the Arctic and the world.
- The UArctic was created based on the initiative of the Arctic Council 20 years ago. Today, the UArctic together with the International Arctic Science Committee (IASC) and International Arctic Social Sciences Association (IASSA) constitutes the three organizations representing Arctic science in the Arctic Council.

The North Matters

The Arctic has changed, and that does influence the rest of the world. In a place where fast and widespread climate change is happening in front of our very eyes, perceiving what we see and acting upon it is a tough task that requires local action but even more international cooperation as the root causes are rarely in the North. Catching up with the myriad of changes to natural, social and political systems is a joint responsibility that rests on all of our shoulders – not only on the northerners'.

The UArctic takes a proactive role in promoting a holistic understanding in Arctic research, including the value of traditional knowledge. Actors in the Arctic region have taken the global lead in promoting the understanding of and respect for northern peoples and their knowledge in Arctic science over the last decade. The UArctic, with its strong commitment to the North and northern perspectives, will continue to be a driver in this for years to come.

The Arctic knowledge map still has many white spots, and it does make a difference what kind of knowledge is used when decisions influencing the future of the North are made. Building relevant knowledge and ensuring that northerners are central in this process are core values for the UArctic. The paper by Ulunnguaq Markussen at the end of the book provides an insightful discussion on the very central challenge of how different knowledge system may support, or be a hindrance, for the future development for and by the North.

Finally, I would like to thank the hard work of the host, the UArctic staff, and others who organized the congress and provided the famous Russian hospitality in the fantastic St. Petersburg. I am also grateful to the contributors and editors of this book which nicely sums up the highlights from scientific contributions at the congress. I look forward to seeing you all in the next UArctic Congress in Oulu in the early fall of 2018.

UArctic   Lars Kullerud
Arendal, Norway

Preface

The chapters of this book are derived from the UArctic Congress 2016 science sessions, focusing on themes identified in the report of the International Conference on Arctic Research Planning (ICARP III) that was published in 2015. Themes address the changes and developments as well as the challenges and opportunities that are taking place in today's global world. The Arctic is changing faster than any other region in the world. Its climate is changing in a speed that cannot be found anywhere else, affecting either directly or indirectly to almost everyone and everything. How can the Arctic societies and cultures, ecosystems, and environments cope with these fast changes?

This book is divided into six thematic parts reflecting the congress themes: Vulnerability of the Arctic Environments, Vulnerability of the Arctic Societies, Building the Long-Term Human Capacity, Arctic Safety, and Arctic Tourism. The final part of the book "Circumpolar, Inclusive and Reciprocal Arctic" looks at the Arctic in the light of the UArctic's mission and values; Gunhild Hoogensen Gjørv, professor of political sciences, addresses a number of issues surrounding the implementation of gender perspectives in the Arctic research, and Ulunnguaq Markussen, UArctic Student Ambassador, calls for an Arctic awakening of peoples in the era when Arctic is seen as a place for natural resource extraction and economic benefits.

During the edition process of the book, it became clear that the chapters represent a cross section of several issues and trends that are currently taking place in the Arctic: increased tourism – also in the seas - calling for maritime safety actions, and preparedness, increasing amounts of contaminants and pollutants in spite of global actions, emerging invasive species that are threatening arctic biodiversity. Changes in the Arctic environment and atmosphere and increasing industrial activities and natural resource extraction affect the Arctic peoples, both indigenous and nonindigenous and both in rural communities and in cities.

Despite the varying topics and different disciplinary approaches, the articles highlight the interconnectedness of the Arctic. Different issues are linked to each other, overlapping, entwined, and scaled across from local to global. The Arctic is not a remote and isolated area but also an area of metropolitan development and a place where global trends are assimilated to, and on the other hand affected by, the Arctic inhabitants.

Making the voice of the Arctic heard in the globalizing world is in the core of the UArctic ambition, reflecting the common values and interests across northern peoples and cultures. We hope that this book succeeds to serve the same purpose and that you enjoy this written journey to The Interconnected Arctic.

Oulu, Finland   Kirsi Latola
                                                                                    Hannele Savela

Contents

Part VI Circumpolar, Inclusive and Reciprocal Arctic

Part I
Vulnerability of the Arctic Environments

Chapter 1
Mysteries of the Geological History of the Cenozoic Arctic Ocean Sea Ice Cover

Jörn Thiede

Abstract The University of the Arctic assembles a large group of northern hemisphere scientific institutions with a huge research capacity due to being the home of a large number of junior scientists with a high potential for the future. The vagueries of the Cenozoic Arctic ice cover history have the potential of contributing to our understanding of future environments on the northern hemisphere. This may have its implications of the socio-economic conditions for the societies inhabiting high northern latitude land areas. Climatic conditions during the young geologic past were sometimes warmer than today; the climate has a „memory" and such conditions might offer analogues what is in store for the future for all of us.

1.1 Introduction

The University of the Arctic Congress during September 2016 probably represented the largest assembly of Arctic research institutions, with a huge potential of many junior scientists who would be able – if combining and coordinating their efforts- to contribute to and to tackle a big problem, namely the variable history of the Arctic Ocean ice cover. Even though the mid-nineteenth-century scientists speculated that the modern central Arctic Ocean may be ice-free, we know since F. Nansen's famous crossing of it on his research vessel FRAM 1893–1896 that it is presently almost completely covered by sea ice, with very few larger ice bergs coming from glaciers and ice shelves from the surrounding glaciated shelves and mountain ranges. The past years of intensive stratigraphic studies of sediment cores from the Arctic Ocean have revealed many new data of changes of the Cenozoic Arctic Ocean ice cover through time. Our understanding of its increasingly complex history is growing and contradicts established text-book knowledge in many ways. This article offers to formulate a great scientific challenge; however, it does not offer for an in-depth going synthesis of this exciting time period of global paleoenvironmental history.

J. Thiede (✉)
Köppen-Laboratory, Institute of Earth Sciences, Saint Petersburg State University,
V. O., Sredniy prospect 41, St. Petersburg 199 178, Russia
e-mail: jthiede@geomar.de

© The Author(s) 2017
K. Latola, H. Savela (eds.), *The Interconnected Arctic — UArctic Congress 2016*,
Springer Polar Sciences, DOI 10.1007/978-3-319-57532-2_1

Fig. 1.1 The Eurasian Arctic shelf seas with their important influx of fresh water are considered the „factories" of the young sea ice, sometimes with large amounts of fine-grained sediments forming „dirty" flows which included into the Transpolar Drift (arrow in the inserted map) travels across the Arctic Ocean to exit mostly into the Norwegian-Greenland Sea through Fram Strait (Photo by H. Kassens, Kiel). The insert map has been adapted from Gorshkov (1983)

1.2   Dynamics of the Modern Arctic Ocean Sea Ice Cover

The Arctic sea ice cover (Fig. 1.1) owes its existence to the polar position of the Arctic Ocean with its small solar insolation and hence cold temperatures, as well as its interaction of the continental hinterlands with the ocean. From modern observations we know that the Arctic sea ice cover is presently shrinking under the influence of rising global temperatures; the year 2016 was the warmest one during the recent past and we will probably experience increasingly open waters in the central Arctic Ocean in the near future.

Its history remained almost completely unknown until the modern research ice breakers were capable to visit the central Arctic Ocean, which for the first time was demonstrated by the Swedish YMER-80 expedition (Schytt 1983). In 1987 the German POLARSTERN reached the Gakkel Ridge; in 1991 POLARSTERN and the Swedish ODEN succeeded as the first conventional surface research vessels to attain the North Pole (Fig. 1.2).

The hydrographic properties of the ocean waters and the glaciology of the ice cover could be studied in great detail. It is now clear that the influx of fresh water from the North American and in particular the Eurasian rivers have a major impact on the formation of a shallow brackish water layer on the surface of the Arctic Ocean which supports the sea-ice cover (Fig. 1.3). Because of the peculiar modern

Fig. 1.2 The International ARCTIC Expedition of the Swedish ODEN and the German POLARSTERN reached the North Pole in early September 1991. It demonstrated that the modern infrastructure now available to polar researchers opened up a completely new phase of Arctic Ocean research (Photo by an expedition participant)

drift pattern of the Arctic sea ice (Transpolar Drift –Fig. 1.1, Beaufort Gyre) the Eurasian shelf seas can considered as „factories for the production" of most of the new sea ice. The formation of this ice can lead to the inclusion of substantial amounts of fine grained sediment materials (Fig. 1.1) which after several melting cycles can concentrate on the ice surface to form „dirty", mostly relatively old ice flows.

1.3 High Variability of Arctic Ocean Ice Covers During the Quaternary

Maps like those published by Hughes et al. (1977) and CLIMAP (CLIMAP Project Members 1976, see also Cline and Hays 1976) provided widely differing opinions of the nature of the Arctic ocean ice cover without having any field data in support. Figure 1.4 assumes that a huge ice sheet extending from North America and northern Eurasia reached across the entire Arctic Ocean forming a thick ice shelf over the deep-sea regions. The CLIMAP reconstruction (Fig. 1.5), on the contrary, saw the ice sheets both on the North American as well as on the NW Eurasian side limited to the continents and their shelves, with the central Arctic Ocean and the Norwegian Greenland Sea possibly covered by sea ice.

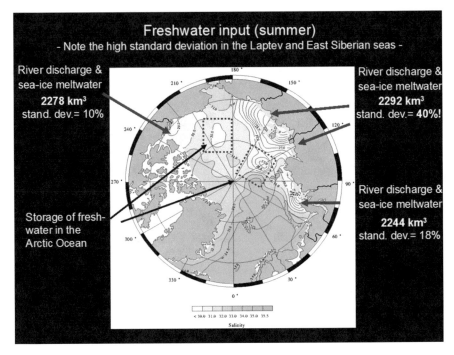

Fig. 1.3 The salinities of the Arctic Ocean surface waters are heavily impacted by the fresh water influx from northern Eurasia. The regional salimity anomalies (isolines are expressed in ppt's) during the summer have been taken from Gorshkov 1983, the corresponding fresh water input has been adapted from EWG (Environmental Working Group) 1998 (Oceanography atlas for the summer period)

Since 1987, research ice breakers have been able to penetrate the Arctic ice cover up to the North Pole, both from the West (Fram Strait and Svalbard) as well as from the Far East (Bering Strait) systematically recovering sediment samples and cores which then allowed to define the impact of the existence of the Arctic ice covers. Based on a collection of sediment cores Spielhagen et al. 2004 have been able to document the presence of the sea ice for the past 120 000 years (MIS 1-5), while the intensity of the sedimentation of planktonic foraminifers as well as of the ice-rafted terrigeneous detritus reflected the alterations between glaciations and interglaciations on the adjacent continents. Each time, one of the large glacial ice sheets collapsed during deglaciation, large amounts of fresh water entered the Arctic Ocean and left their imprint in the O-isotope signals preserved in the shell materials of the planktonic foraminifers. Mangerud et al. 2004 also studied the impact of the subglacial lakes whose run-off to the Arctic Ocean had been barred by the glacial ice sheets in North America and NW Eurasia while rivers such as the Lena and others to the East of Taymyr Peninsula (eastern limitation of the Upper Quaternary ice sheets) had been emptying their fresh water into the Arctic Ocean almost continuously during glacials and interglacials.

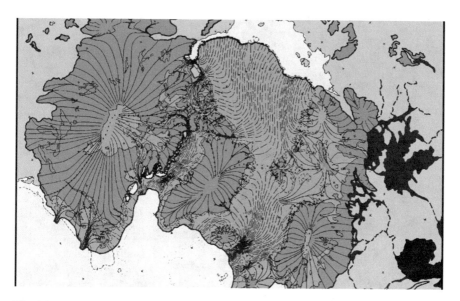

Fig. 1.4 Northern Hemisphere ice cover during the LGM (Last Glacial Maximum) according to Hughes et al. (1977) who assumed the existence of a thick ice shelf over the central Arctic deep-sea basins

This simplified picture changed substantially when the traces of large ice shelves extending into the central Arctic Ocean were detected through glacial erosional features, first on Lomonosov Ridge, later also on several other structural highs in the Arctic Ocean (Jakobsson et al. 2016). This interpretation is based on geophysical evidence from multibeam bathymetry and seismic reflection subbottom profiles. The ice shelves appear to be coeval with MIS 6 which represented the glaciation over NW Eurasia with an ice sheet substantially larger than the younger ice sheets which had been mapped by Svendsen et al. 2004 (cf. Thiede et al. 2004) as their contribution to the QUEEN project.

It would be highly desirable to learn more about the Cenozoic, in particular the early Pleistocene history of the Siberian river run-off to the Arctic Ocean, because of its influence on the Arctic Ocean surface water salinities. There are well developed river terrace systems from East Siberian rivers such as the Lena and it is hoped that they will be studied intensely in the future (Savelieva et al. 2013). The Quaternary history of this region and of the bed of Lena river in this region is poorly known, but of particular interest, since an extension of the Russian railroad system has been established along the eastern shores of the Lena River; it is probably in the vicinity of Yakutsk (capital of the republic of Sakha) where the first bridge ever will cross the most important, longest and undisturbed Siberian river draining into the Arctic Ocean.

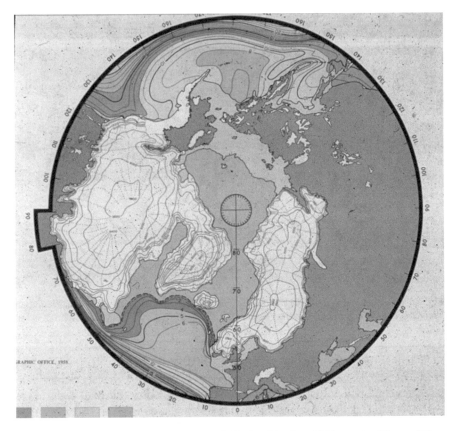

Fig. 1.5 The CLIMAP reconstruction (CLIMAP Project Members 1976; see also Cline and Hays 1976) of the last glacial maximum (LGM) on the Northern Hemisphere. Contrary to Hughes et al. (1977) the central Arctic Ocean was considered sea ice covered, even this could only be proven much later after expeditions of research ice breakers successfully reached the central Arctic Ocean (Spielhagen et al. 2004)

1.4 The Enigma of the Tertiary Predecessors of the Central Arctic Ocean Ice Covers

Already Köppen and Wegener (1924) based on observations from northernmost Siberia had vaguely assumed that Cenozoic northern hemisphere glaciation reached back into the Miocene. This assumption could be confirmed by the activities of the deep-sea drilling project in the Norwegian-Greenland Sea (DSDP/ODP Legs 38, 104 and 151), the two latter legs recovering completely cored Lower and Upper Tertiary sediment sequences (Thiede 2011a, b). In the Norwegian-Greenland Sea and in the drill sites around Greenland indications of continuous ice-rafting of coarse terrigenous sediment components suggested that ice-bergs were present in this area since Paleogene times which while melting shed their sedimentary load to the seafloor.

The JOIDES RESOLUTION was the first scientific drilling vessel which succeeded –accompanied by the heavy duty ice-breaker FENNICA- to enter the Arctic Ocean and recover sediment cores from the Neogene and Pleistocene sediment cover of the Yermak Plateau to the North of Svalbard (Thiede, Myhre et al. 1996). A most spectacular record of ice-rafted fine- and coarse grained terrigeous has been obtained by the IODP Expedition 302 ARCTIC CORING (ACEX) of ECORD in 2004 when a flotilla of three ice breakers reached a position on Lomonosov Ridge very close to the North Pole (Backman and Moran 2009). Figure 1.6 is reflecting the analyses of the shipboard sedimentologist St. John (2008) with observations af variable concentrations of coarse and fine-grained ice rafted detritus since Late Eocene times. The precise age of the onset of ice-rafting in the Upper Eocene sediments is still a subject of debate (Poirier and Hillaire-Marcel 2011) but it is clear that this happened much earlier than assumed hitherto (apparently also earlier than the onset of Antarctic glaciation). A similar very early onset was later confirmed at ODP Site 913 immediately to the East of Greenland.

The presence of the coarse grained ice-rafted sediment components suggest the occurrence of ice-bergs originating from glaciers or ice-shelves which reached into the ocean. This cannot be said to easily for the finegrained materials which can also

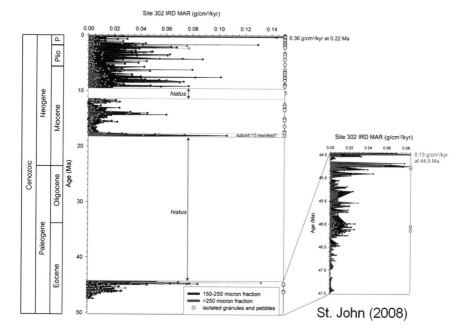

Fig. 1.6 The Lomonosov Ridge ice-rafting record from Eocene through Quaternary times deduced from sediment components (St. John 2008). The *yellow circles* mark the occurrence of relatively coarse terrigenous deposits indicative of ice bergs. Age scales on the *left*; a hiatus marks a missing sediment record

be brought in by sea ice. Other researchers have therefor argued, based on proxies such as IP_{25} for spring sea ice cover and alkenone-based summer sea-surface temperatures > 4 °C that the central Arctic was only seasonally ice-covered (Stein et al. 2016) during the Late Miocene.

The same applies to Upper Cretaceous biosiliceous laminated sediments which had been collected many years ago from Alpha Ridge. Their microfossils seemed to indicate summer sea surface temperatures of 10–15 °C, but later detailed analysis of the laminations documented the presence of thin sand layers in between suggesting ice-rafting by winter sea ice and hence documenting a strong seasonal difference (Davies et al. 2009).

1.5   What Triggered the Onset of Northern Hemisphere Glaciations During the Paleogene?

This question leaves room for speculations, but the idea of seeking a relationship between tectonism to the origin and history of continental drainage systems and Arctic climate is by no means new (pers. comm. Hayes 1996; Molnar and Tapponier 1975; Ruddiman & Kutzbach 1989) even though the available stratigraphies and the timing of events have completely changed over the past 20 years.

Brinkhuis et al. (2006) have published a paleogeographic map for the Eocene Arctic Ocean, when the occurrence of *Azolla*-microspores in the ACEX-drill cores from Lomonosov Ridge suggested substantially warmer temperatures than later. In modern times *Azolla* is a floating fresh water fern, mostly lives in subtropical fresh water environments.

During the Eocene, the Arctic Ocean was connected to the world ocean only through shallow sea ways. The Turgay Strait across western Siberia linked the Arctic Ocean with the Tethyan system further to the South suggesting that the Siberian platform at that time was not tipping to the North. *Azolla* has also been observed in Eocene drill cores in the Norwegian-Greenland and Labrador seas, apparently recording a very strange event of large amounts of fresh water entering the high northern latitude seas.

The deposition of the *Azolla*-rich sediments preceded the onset of ice-rafting in the ACEX-cores and it is tempting to link the onset of northern hemisphere glaciation to the plate tectonic processes occurring at the southern margin of the Eurasian Plate resulting in the generation of a northward flowing drainage system emptying into the Arctic Ocean (Fig. 1.7).

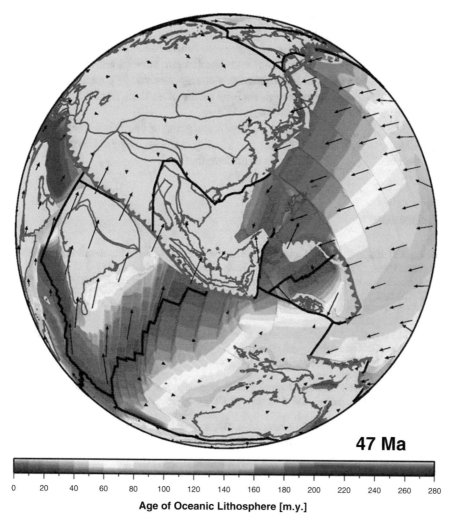

47 Ma

0   20   40   60   80   100   120   140   160   180   200   220   240   260   280

Age of Oceanic Lithosphere [m.y.]

Fig. 1.7 The late Mesozoic and early Cenozoic plate tectonic development resulted in the collision of the Indian subcontinent with the Southern Eurasian plate margin with important consequences for the physiography of the Siberian platform which developed since that time a river system almost exclusively draining into the Arctic Ocean (figure via personal communication of D. Müller, Sydney). The age distribution of the deep ocean floors can be deduced from the distribution of well-dated magnetic anomalies which have been mapped in the global ocean and which allow to define the plate tectonic evolution

1.6 Conclusions

No question: The Late Mesozoic and Early Tertiary Arctic Ocean was ice-free at times, despite the occurrence of „glendonites" (mineral formations indicative of cold temperatures) in Upper Cretaceous and Paleogene sediments around the Arctic.

Nobody knows where and precisely when or why the onset of Cenozoic Northern Hemisphere glaciation occurred, but certainly much earlier that believed up to now! Urgent research needs comprise of systematic future Arctic deep-sea drilling with good areal coverage based on solid and detailed site surveys and a resolution of Siberian river histories, and also involve a courageous young generation of polar geoscientists because to resolve this exciting period of northern hemisphere paleoenvironmental history will take decades.

The University of the Arctic is the largest gathering of young polar scientists and provides the logical and best platform to identify motivated, internationally oriented and well educated junior scientists to achieve this goal.

References

Backman J, Moran K (2009) Expanding the Cenozoic paleoceanographic record in the Central Arctic Ocean. IODP Expedition 302 synthesis. Centr Europ J Geosci 1:157–175

Brinkhuis H, Schouten S, Collinson ME et al (19 add. co-authors and Expedition 302 Scientists) (2006) Episodic fresh surface waters in the Eocene Arctic Ocean. Nature 441: 606–609

CLIMAP Project Members (1976) The surface of the ice-age Earth. Science 191(4232):1131–1137

Cline RM, Hays JD (eds) (1976) Investigation of Late Quaternary paöepceanography and paleoclimatology. Geol Soc Amer Mem 145:464 (Boulder, CO)

Davies A, Kemp AES, Pike J (2009) Late Cretaceous seasonal ocean variability from the Arctic. Nature 460:254–258

EWG (Environmental Working Group) (1998) Oceanography atlas for the summer period. University of Colorado, Boulder, version 1.0. CD from User Services NSIDC/CIRES, Boulder, CO

Gorshkov SG (ed) (1983) World Ocean Atlas. Vol. 3. Arctic Ocean.- 184 pp., (Pergamon Press) (copyright Dept. Navig. Oceanogr., USSR Ministry of Defense) Leningrad (in Russian with annotation in English)

Hughes TJ, Denton GH, Grosswald MG (1977) Was there a late Würm Arctic ice sheet? Nature 266:596–602

Jakobsson M, Nilsson J, Anderson L et al (22 add. co-authors) (2016) Evidence for an ice shelf covering the central Arctic Ocdan during the penultimate glaciation. Nat Commun 7: 10.1038/ncomms10365

Köppen W, Wegener A (1924) Die Klimate der Geologischen Vorzeit/the climates of the geological past. In: Thiede J, Lochte K, Dummermuth A (eds) Introduction by the editors, reprint and new translation into English. Borntraeger Science Publishers, 657 S

Mangerud J, Jakobsson M, Alexanderson H et al (11 add. co-authors) (2004) Ice dammed lakes and rerouting of northern Eurasian drainage duriing the last glaciation. Quat Sci Rev 23: 1313–1332

Molnar P, Tapponier P (1975) Cenozoic tectonics of Asia: effects of a continental collision. Science 189:419–426

Poirier A, Hillaire-Marcel C (2011) Improved Os-isotope stratigraphy of the Arctic Ocean. Geophys Res Let 38:L14607. doi:10.1029/2011GL047653

Ruddiman WF, Kutzbach JE (1989) Forcing of late Cenozoic Northern Hemisphere climate by plateau uplift in southern Asia and American West. J Geophys Res D15:18409–18427

Savelieva L, Bolshyanov D, Thiede J (2013) Среднее течение р. Лена – Нижний Бестях – Майя – Эдейцы – Бестях.- (LENA-Expedition 2013).- Report Köppen-Laboratory SPbGU, 15 pp., 8 .figs., Saint Petersburg (in Russian)

Schytt V (1983) YMER-80: a Swedish expedition to the Arctic Ocean. Geogr J 149(1):22–28

Spielhagen RF, Baumann KH, Erlenkeuser H et al (4 add. co-authors) (2004) Arctic Ocean deep-sea record of northern Eurasia ice sheet history. Quat Sci Rev 23: 1455–1483.

St. John K (2008) Cenozoic ice-rafting history of the central Arctic Ocean: terrigenous sands on the Lomonosov Ridge, -Paleoceanography, 23, PA1S05, doi:10.1029/2007PA001483

Stein R, Fahl K, Schreck M (10 add. co-authors) (2016) Evidence for ice-free summers in the late Miocene central Arctic Ocean. Nat Commun 7(Art. nr: 11148): doi:10.1038

Svendsen JI, Alexanderson H, Astakhov V et al (27 add. co-authors) (2004) Late Quaternary ice sheet history of northern Eurasia. Quat Sci Rev 23:1229–1271

Thiede J, Myhre AM, Firth JV, Shipboard Scientific Party (1996) Proceedings ODP, Scientific Results 151: 685 pp., (Ocean Drilling Program) College Station, TX

Thiede J, Astakhov V, Bauch H et al (9 add. co-authors) (2004) What was QUEEN? Its history and international framework. Quat Sci Rev 23:1225-1227.

Thiede J, Jessen C, Kuijpers A, Mikkelsen N, Nørgaard-Pedersen N, Spielhagen RF (2011a) Million years of Greenland Ice Sheet history recorded in ocean sediments. Polarforschung 80(3):141–159

Thiede J, Eldholm O, Myhre A (2011b) Deep-sea drilling in the Norwegian-Greenland Sea and Arctic Ocean. In: Spencer AM, Gautier D, Stoupakova A, Embry A, Soerensen K (eds) Arctic petroleum geology. Geological Society, London, pp 703–714

Chapter 2
Response of Arctic Alpine Biota to Climate Change – Evidence from Polar Urals GLORIA Summits

Yuri E. Mikhailov and Pavel A. Moiseev

Abstract Polar Urals as one of target regions of Global Observation Research Initiative in Alpine Environments (GLORIA) comprises a suite of four summits, representing an elevation gradient of alpine vegetation patterns. The sampling areas cover the summits from the tops down to the 10 m contour line and are divided into eight sections. For each section, a complete list of vascular plants and herpetobiotic arthropods was collected and resurveyed. In the period from 2001 untill 2015, the species numbers steadily increased and the total surplus of vascular plants was up to 13 species on separate summit. A general decrease in the total cover of vascular plants and changes in percentage cover of the dominant species was recorded on the permanent plots; certain species of herbs decreased and certain shrub species increased. Among the dominant species of invertebrates, ground beetles and millipedes were replaced by click beetles and spiders. After 14 years the altitudinal index calculated for vascular plants gave an average upward movement of 13.6 m, that is more pronounced than in Northern and Southern Urals. The thermophilization of the alpine plant communities of Polar Urals was found equal to 9.3% of one vegetation belt. The temperature sums obtained from data loggers demonstrate the slight tendency of increase, especially for the lower summits.

Y.E. Mikhailov (✉)
Ural Federal Yeltzin University, Lenin prosp., 51, Yekaterinburg 620002, Russia

Ural State Forest Engineering University, Sibirsky trakt, 37, Yekaterinburg 620100, Russia
e-mail: yuemikhailov@gmail.com

P.A. Moiseev
Institute of Plant and Animal Ecology, 8 Marta str., 202, Yekaterinburg 620144, Russia
e-mail: moiseev@ipae.uran.ru

K. Latola, H. Savela (eds.), *The Interconnected Arctic — UArctic Congress 2016*,
Springer Polar Sciences, DOI 10.1007/978-3-319-57532-2_2

2.1 Introduction

The field sites of this study are located in the Polar Urals and constitute a part of the international long-term monitoring network GLORIA (Global Observation Research Initiative in Alpine Environments). This GLORIA initiative focuses on the alpine life zone, the area above the forestline, for tracing and understanding the response of alpine ecosystems to ongoing climate change (Pauli et al. 2015). The vascular plant species occurrence was recorded first in the year 2001 on 72 mountain summits distributed across 18 study regions in Europe, with Polar Urals among them. Since then the surveys were repeated twice, in 2008 and 2015.

The Urals is a unique submeridional mountain range in Eurasia, extending over 2000 km in north-south direction from the Barents Sea and Kara Sea shores at ca. 70°N to the Pre-Aral Sea sands at ca. 48°N. The mountain range crosses several zonobiomes from arctic tundra to steppe and is, accordingly, divided into five orographic regions: Polar, Subpolar, Northern, Middle and Southern Urals (Mikhailov and Olschwang 2003). From these regions, Polar Urals is situated mainly inside the North Polar Circle and is a part of the Arctic floristic region as delimited by Yurtsev (1994). Alongside with Khibiny, Putorana Plateau and other meta-arctic mountains (Makarova et al. 2013), Polar Urals is situated at the North Polar Circle and having alpine life zone in closest contact with the arctic tundra zonobiome. That is why the results of this study are of interest for both alpine and Arctic research.

In contrast to meteorological and glaciological studies, long-term observations of climate change impacts on alpine ecosystems are scarce and mainly based on incidental historical data from the Alps, Scandes and the Scottish Mountains (Pauli et al. 2015). For the arthropod communities similar records are limited mainly for 20–30 years comparison of the ground beetle assemblages in the Apennines and the Dolomites (Brandmayr et al. 2002; Pizzolotto et al. 2014).

Resurveys of the historic summit sites in the Alps showed that vascular plants have been found at higher altitudes than recorded earlier (Pauli et al. 2012). Furthermore, recent resurveys of 60 summits of the major European mountain ranges provide evidence that more warm-demanding plant species, which usually dwell at lower elevations, increase and more cold-adapted species occurring at high elevations show concurrent decline. This process was described as thermophilization of alpine plant communities (Gottfried et al. 2012). Alongside with it, a general upward shift of plant species was observed across the continent during the past decade. In northern and central Europe, this led to an increase in species numbers, while in the Mediterranean region, species numbers were stagnating or decreasing (Pauli et al. 2012).

Within alpine biota only very few attempts have been made to search for a restricted set of organisms acting as surrogates for overall species richness. According to Pauli et al. (2015) vascular plants as sessile and macroscopic organisms that can be readily identified in the field represent such a group. On the other hand, the studies of Finch and Löffler (2010) in alpine areas of Norway resulted in advice to use, apart from vegetation, at least one abundant group of invertebrates as indicator for general species richness pattern. The arthropods that depend strongly on abiotic factors and plant species distribution are exactly such a group.

2.2 Materials and Methods

The detailed description of the standard long-term monitoring design and method of GLORIA, is given in the field manual available online (Pauli et al. 2015), where the authors of this paper contributed as well. According to this Multi-Summit Approach a target region in the Polar Urals was established in 2001. It comprises a suite of four summits (see Table 2.1) distributed in equal elevation intervals and representing an elevation gradient of vegetation patterns, characteristic for this mountain region (Gorchakovky 1975). The summits are situated in Yamalo-Nenets region of Russia in the valley of Sob' river ca. 45 km NNW Salekhard (66°54′ – 67°00′ N, 65°35′ – 65°46′ E) along the Sob'-Elets passageway and quite easily accessible by the railway from Labytnangi to Ust-Vorkuta.

The sampling area covered each summit from its top (highest summit point) down to the 10 m contour line and was divided into eight summit area sections (SAS). The standard records for each SAS include a complete species list with the estimation of the abundance of each species and percentage top cover of surface types. The 3m×3m quadrat clusters are placed at the 5-m level in all four main compass directions (Fig. 2.1).

Each quadrat cluster includes four 1-m^2 quadrats, where the top cover of surface types and cover of each vascular plant species are recorded (Pauli et al. 2015). Four T-loggers (GeoPrecision Mlog-5W) positioned on each summit has been measuring the soil temperature 10 cm below the surface at hourly intervals since 2001 until 2015. Polar Urals is one of the few GLORIA target regions, where invertebrate monitoring as an extra approach is used according to research protocol by one of the authors (Mikhailov 2015).

The field teams included leading experts of regional biodiversity, able to identify all vascular plant taxa at the vegetative stage on site. For arthropods the exact determination of collected specimens followed later on by comparing with reference

Table 2.1 The number of species of vascular plants found on the summits of Polar Urals during the three monitoring campaigns

Summit	Altitude, m a.s.l.	Raw data			Filtered data		
		2001	2008	2015	2001	2008	2015
Pourkeu (POU)	839	4	3	3	0	0	0
Malyi Pourkeu (MPO)	641	51	54	61	46	47	51
Slantsevaya (SLA)	417	38	45	51	34	39	40
Shlem (SHL)	300	35	38	37	26	28	27
Total in target region		72	79	90			

Raw data columns refer to unprocessed data, filtered data columns to the data pre-processed

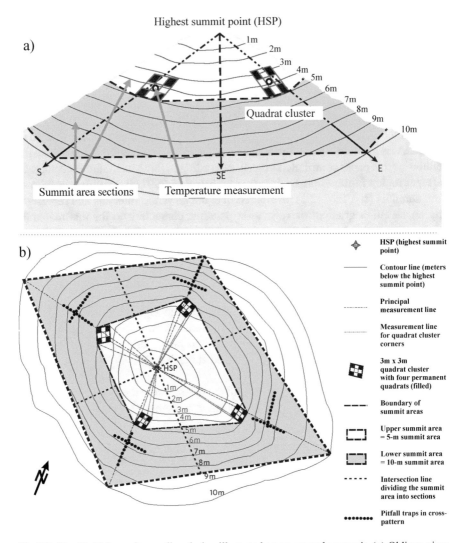

Fig 2.1 The Multi-Summit sampling design illustrated on an example summit. (**a**) Oblique view with schematic contour lines; (**b**) Top view. The 3m×3m quadrat clusters, pitfall traps in cross-pattern (extra approach) and the corner points of the summit areas are arranged in the main geographical directions (Modified from Pauli et al. 2015)

collections or by specific expertise. To gain representative data the results from single pitfall traps were pooled for each line of ten traps in each SAS; for plants data from single SAS and 1-m² quadrat were used in the analyses as independent replicates. To avoid observation errors in collecting species lists the same experts participated in all surveys and the data was filtered to remove single records and potentially misidentified species from the list before analysis.

To evaluate whether the changes in summit species richness might be related to a possible upward or downward move of species ranges, an altitudinal index for each species of vascular plants and first for invertebrates was calculated following Pauli et al. (2012).

For detection of a warming effect or thermophilization the first step is assigning an altitudinal rank to all recorded plant species according to their optimum performance (see Supplementary Materials for Gottfried et al. (2012)). Using data on altitudinal ranks and cover of each plant species within 1-m^2 quadrats a composite score in the following thermic vegetation indicator S was then calculated. Its differences between respective years were used to quantify transformations of the plant communities and termed thermophilization indicator D ($D = S_{year1} - S_{year2}$) (Gottfried et al. 2012).

2.3   Results and Discussion

Using data from T-loggers we computed temperature sums that influence species diversity patterns. As the threshold temperatures for plant growth in alpine regions are not commonly agreed (Körner 2012), we used temperature sums above 3 °C for May–July and from them revealed slight tendency of increase especially for the lower summits (SLA and SHL) (Fig. 2.2).

On two summits with polygonal alpine tundra (SLA and MPO), the number of plant species has been steadily increasing as estimated in summit area sections. Between 2001 and 2008 the increase was from 3 to 7 species, between 2008 and 2015 – from 6 to 7 species (Table 2.1). After 14 years the surplus was up to 13 species on the separate summit and 18 species for the whole target region.

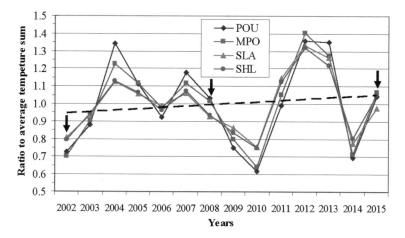

Fig. 2.2 Temperature sums > 3 °C for May-July (indexed to the average value) computed from T-loggers data from the summits of Polar Urals

Across 66 summits surveyed by GLORIA-Europe, the average numbers of species per summit increased by ca. 8% between 2001 and 2008. However, only in boreal and temperate regions most summits have gained additional species (average increase was from 38.0 to 41.9, that is 10.3%), while in Mediterranean regions the majority of summits had less species (Pauli et al. 2012). In the Polar Urals there was a similar 10% increase for this period and 10.6% between 2008 and 2015 (Table 2.1).

Comparable data for herpetobiotic arthropods is available between 2008 and 2015 for two lower summits (SHL and SLA) where the increase in number of species yielded 3–4 for all species and 2–4 after filtration of species lists.

At the continental scale, counts of altitudinal index from 2001 till 2008 suggest that plant species were shifting their distributions to higher altitudes by 2.7 m on average (Pauli et al. 2012). The results for the Polar Urals for the same period gave the shift for 9.2 m and from 2008 till 2015 – for 4.4 m so that in general for 14 years the shift was 13.6 m upslope. In opposite, for invertebrates, the shift was 20.6 m downslope from 2008 to 2015.

With the general decrease of total cover of vascular plants on the permanent quadrat clusters, changes were found in percentage cover of the dominant species in the communities between the years 2001 and 2015. The common tendency was a decrease of some species of herbs (e.g. *Luzula confusa* Lindeb., *Festuca ovina* L.) and increase of certain shrub species (e.g. *Ledum decumbens* (Ait.) Lodd. ex Steud, *Dryas octopetala* L., *Vaccinium vitis-idaea* L.). The percentage cover of alpine sweetgrass *Hierochloa alpina* (Willd.) Roem. & Schult., a common species on all summits, was obviously decreasing on eastern and northern sectors and increasing in southern sectors.

The dominance level for each arthropod species was calculated according to the classification of Engelmann (1978) and the dominance structure was analyzed at species and family levels. The most comparable results during surveys in 2008 and 2015 were obtained for SLA summit (Table 2.2).

In both surveys true bugs *Chlamydatus opacus* Zett. and *Calacanthia trybomi* Sahlb. were dominating, but in 2008 they were co-dominating while in 2015 *C. trybomi* got higher dominance class than *Ch. opacus* and the percentage almost five times as higher. Besides, the most notable tenfold decrease of abundance was found for ground beetle *Carabus truncaticollis* Popp. while click beetle *Oedostethus similarius* Dolin et Medv. increased its abundance also tenfold and Lycosid spider *Pardosa septentrionalis* West. increased four times.

For the Polar Urals between 2001 and 2008, the thermophilization indicator D was 0.057, almost the same value as a European mean (Gottfried et al. 2012). In the next period, between 2008 and 2015, D was lower (0.037) (Table 2.3).

This gives the extent of transformation of alpine summit communities in the Polar Urals in the magnitude of 5.7% of one vegetation belt between 2001 and 2008, 3.7% between 2008 and 2015 and resulting 9.3% for 14 years. Among the summits, the third one, MPO, had almost no changes, the lowest one (SHL) represented almost the same D values as the average for the region, while the second summit, SLA, had the highest value of 18% of one vegetation belt for 14 years (Table 2.3).

Table 2.2 The dominance structure of herpetobiotic arthropods on Slantsevaya (SLA) summit at species and family levels

Dominance class	2008		2015	
	Species	Family	Species	Family
Eudominant	Chlamydatus opacus	Miridae	Calacanthia trybomi	Saldidae
	Calacanthia trybomi	Saldidae		
Dominant	Lithobius curtipes	Lithobiidae	Chlamydatus opacus	Miridae
Subdominant	Carabus truncaticollis	Carabidae	Alopecosa hirtipes	Lycosidae
	Alopecosa hirtipes	Lycosidae	Pardosa septentrionalis	Lycosidae
			Oedostethus similarius	Elateridae
S (number of species)	21		24	
N (total number of collected specimens)	469		410	
H (Shannon index)	1.79 ± 0.05		1.71 ± 0.07	
E (evenness index)	0.59		0.54	

Table 2.3 Thermic vegetation indicator (S) and thermophilization indicator (D) for the summits of Polar Urals for each period of survey

	SHL	SLA	MPO	D average
S 2001	3.36	3.40	3.00	
S 2008	3.41	3.51	3.01	
S 2015	3.45	3.58	3.01	
D 2001–2008	0.05	0.11	0.01	0.057
D 2008–2015	0.04	0.07	0.00	0.037
D 2001–2015	0.09	0.18	0.01	0.093

The presented data on climate-induced transformation of alpine biota within 14 years can be considered a rapid ecosystem response to ongoing climate change. In comparison with Northern and Southern Urals upwards migration of more warm-demanding plant species in the Polar Urals is more pronounced. The lack of similar data from other meta-arctic and arctic mountains prevents us from making broader circumpolar conclusions, however ongoing GLORIA activities expect further data from Alaska, Greenland and Svalbard.

Acknowledgments We are grateful to Dmitry Moiseev and Alexander Ermakov for their valuable help during the fieldwork. Thanks to the funding from Russian Foundation for Basic Research (RFBR grant 15-05-05549 A) the recent resurvey, congress participation and data analysis became possible. The initial species and temperature data recording was supported through the EU FP-5 project GLORIA-Europe.

References

Brandmayr P, Zetto T, Colombetta G, Mazzei A, Scalercio S, Pizzolotto R (2002) I Coleotteri Carabidi come indicatori predittivi dei cambiamenti dell'ambiente: clima e disturbo antropico. In: Atti XIX Congresso nazionale italiano di Entomologia Catania 10-15 giugno 2002, pp 283–295

Engelmann H-D (1978) Zur Dominanz Klassifizierung von Bodenartropoden. Pedobiologica 18:378–380

Finch O-D, Löffler J (2010) Indicators of species richness at the local scale in an alpine region: a comparative approach between plants and invertebrate taxa. Biodivers Conserv 19:1341–1352

Gorchakovsky PL (1975) Rastitelnyi mir vysokogornogo Urala (Flora of high mountains of the Urals). Nauka, Moscow

Gottfried M, Pauli H, Futschik A et al (2012) Continent-wide response of mountain vegetation to climate change. Nat Clim Chang 2:111–115

Körner C (2012) Alpine treelines: functional ecology of the global high elevation tree limits. Springer, Basel

Makarova OL, Makarov KV, Berman DI (2013) Zhuzhelitsy (Coleoptera, Carabidae) vysokogoriy Olskogo Plato, Kolymskoye Nagorye (Ground beetles (Coleoptera, Carabidae) of the Ola Plateau Highlands, Kolyma Uplands). Zoologichesky Zhurnal 92(8):927–934

Mikhailov Y (2015) Invertebrate monitoring on GLORIA summits. In: Pauli H, Gottfried M, Lamprecht A et al (eds) The GLORIA field manual – standard Multi-Summit approach, supplementary methods and extra approaches, 5th edn. GLORIA-Coordination, Vienna, pp 70–71. http://www.gloria.ac.at/methods_manual.html. Accessed 20 Dec 2016

Mikhailov YE, Olschwang VN (2003) High altitude invertebrate diversity in the Ural Mountains. In: Nagy L et al (eds) Alpine biodiversity in Europe, Ecological Studies, vol 167. Springer, Heidelberg, pp 259–279

Pauli H, Gottfried M, Dullinger S et al (2012) Recent plant diversity changes on Europe's mountain summits. Science 336:353–355. doi:10.1126/science.1219033

Pauli H, Gottfried M, Lamprecht A et al (eds) (2015) The GLORIA field manual – standard Multi-Summit approach, supplementary methods and extra approaches, 5th edn. Vienna, GLORIA-Coordination, Austrian Academy of Sciences & University of Natural Resources and Life Sciences. http://www.gloria.ac.at/methods_manual.html. Accessed 20 Dec 2016

Pizzolotto R, Gobbi M, Brandmayr P (2014) Changes in ground beetle assemblages above and below the treeline of the Dolomites after almost 30 years (1980/2009). Ecol Evol 3:1–11. doi:10.1002/ece3.927

Yurtsev BA (1994) Floristic division of the Arctic. J Veg Sci 5(6):765–776

Chapter 3
The Features of Natural and Artificial Recovery in Quarries of the Forest-Tundra Zone of Western Siberia

Elena Koptseva and Alexander Egorov

Abstract The features of natural recovery and artificial restoration of quarries in northern regions of Western Siberia is analysed in this article. The effectiveness of restoration is compared with natural revegetation results. It is shown that the development of restoration projects requires a comprehensive consideration of environmental conditions. General recommendations for restoration of quarry areas are formulated.

3.1 Introduction

Quarry is a common type of land-use in northern regions of Russia. A quarry is an open-cut excavation that reaches over a considerable transversal size. It is used for extraction of sand, loam, stone and others. Post-operative quarry area remains without vegetation and fertile soil cover for a long time. Environmental conditions in quarries are very specific for plants, because soil is nutrient poor and erosion processes are very active (Abakumov et al. 2011). Usually, quarry area has a complex internal anthropogenic relief in the form of high peaks, pits and furrows.

Publication activity on the issue of disturbed lands restoration shows the absence of consensus. What is more effective: natural or artificial recovery? Previous publications indicate that the process of natural revegetation may take many years and depend on many environmental factors (Borgergard 1990; Sumina 1998).

E. Koptseva (✉)
Saint-Petersburg State University, Saint-Petersburg, Russia
e-mail: ekoptseva@hotmail.com

A. Egorov
Saint-Petersburg State University, Saint-Petersburg, Russia

Saint-Petersburg State Forest Technical University, Saint-Petersburg, Russia

© The Author(s) 2017
K. Latola, H. Savela (eds.), *The Interconnected Arctic — UArctic Congress 2016*,
Springer Polar Sciences, DOI 10.1007/978-3-319-57532-2_3

Information about efficiency of artificial land restoration is contradictory and practical approaches used for the restoration are not summarized. In some regions of Russia, for example in the Yamal-Nenets Autonomous District, the regional standards for restoration of disturbed lands are still absent. The nationwide standards have to use but they do not fully take into account peculiarities of northern ecosystems.

Gained actual experience in land restoration, as well the capacity of northern ecosystems to natural recovery should be considered in the development of regional standards. In this context, a comparison of natural recovery effectiveness with artificial restoration of disturbed lands, as well as generalization of accumulated practical experience on restoration, is particularly relevant for the Far North of Russia.

3.2   Materials and Methods

Natural and artificial vegetation recovery on quarries was studied in several geographical locations in the Yamal-Nenets Autonomous District of Northwest Siberia (Fig. 3.1).

The study area is covered by forest-tundra plant communities. Larch, larch-spruce and larch-pine sparse forest and palsa mires are predominated in the region (Fig. 3.2 a, b). Species of shrubs (as *Betula nana*), dwarf shrubs (as *Ledum palustre, Vaccinium myrtillus, V. uliginosum, V. vitis-idaea, Empetrum hermaphroditum*), mosses (as *Pleurozium schreberi, Hylocomium splendens*) and fruticose lichens (as *Cladonia* and *Cetraria* species) have the highest frequency and are much more abundant in undisturbed zonal vegetation.

Natural vegetation recovery as well as artificial processes was investigated from 2013 to 2016 on sandy or sandy loam quarries. Sometimes the substrate contained a small admixture of gravel and pebble. Development of quarries was finished at different times (from a few to 40 and more years ago). The basic method of study was the inventory of vegetation in ten quarries by sample plots from 25 to 400 sq.m. The total number of plots was more than 40. In each sample plot we collected common data on:

- total list of vascular and non-vascular species;
- total vegetation cover (%), projective cover for each of vascular and non-vascular species (%), projective cover (%) for each of the main plant types (woody plants, herbs, forbs, graminoids, shurbs, mosses and lichens);
- the number and the height of layers of vegetation.

Community types were distinguished according to Cajander's theory of forest types by taking into account the combination of dominant species (Barrington 1927).

Granulometric analysis of soil substrates was performed according to the method of sifting cylinder (Soil Sampling and Methods of Analysis 2008).

Fig. 3.1 The map of the Yamal-Nenets Autonomous District. Locations of field data collection sites are indicated by *red marks*

3.2.1   *Natural Recovery of Vegetation*

In quarries, natural recovery takes place in several stages. In this study, four successional stages were distinguished. Sparse vegetation varying in the species composition of plants is formed in the quarry during the first few years (Fig. 3.3a). This stage of the recovery can be described as an ecological chaos because ecologically different species coexist in such habitats. As the result, the maximum of plant species diversity is present in this stage. Interactions between plants are extremely weak and instable. Usually such plant communities are named as pioneers (Walker and del Moral 2008). A small proportion of native species (not more than 15%) from undisturbed surrounding plant communities comes to the quarry during first years of the recovery. This is because most of the native plants have competitive or stress tolerant ecological strategies whereas the environmental conditions in quarries are favorable for ruderal species (Forbes et al. 2001). Species such as willow herb, chamomile,

Fig. 3.2 Natural undisturbed sparse forest (**a**) and palsa (**b**) in the region (Photographs by: E. Koptseva)

coltsfoot, knotgrass, horsetail, fescue grass, snow grass are the most typical for this stage on the study areas.

Horizontal and vertical structure of vegetation also has not yet formed at this initial stage (Walker and del Moral 2008). As a general rule, plants varied in height are present in pioneer communities. Horizontally, vegetation begins to form from separate patches as shown in Fig. 3.3a. Patchy vegetation structure is probably due to clonal growth of some plants for example like willow herb or coltsfoot. So, more than 50% of common species in quarries prefer vegetative reproduction. Clonal growth of plants is advantageous in unstable environmental conditions (Witte and Stöcklin 2010).

At this stage, the pioneer communities of annual and perennial herbaceous plants are dominated. This stage usually lasts for several years (not more than 10 years). But pioneer communities can exist on steep erosion slopes even during the whole recovery time of over 40 years (Fig. 3.3b).

Next stage of the recovery is determined by the local environmental factors. Ecological selection leads to the formation of plant species groups with similar ecological requirements. Dominant species begin to stand out depending on soil granulometric composition and moisture conditions. The vegetation structure (especially vertical) begins to form in the community. Usually sparse layers of shrubs, herbaceous plants and mosses are already present. For example, willow thickets in wet quarry bottoms are particularly characteristic for this stage (Fig. 3.3c).

Significant changes in the vegetation cover of the quarry occur during the third stage, when recovering vegetation will begin to influence the environment by transformation of main environmental factors such as moisture, light, nutrients and others. For example, the growth of moss cover can keep soil moisture very effective. So, vegetation begins to create special micro environment. Species composition of the community is already similar to those in undisturbed areas. Ruderal species and apophytes have a low frequency and abundance. The features of vertical and horizontal

Fig. 3.3a The first stage of the vegetation in recovering quarry, i.e. "ecological chaos"

Fig. 3.3b Sparse pioneer vegetation on steep slopes in "old" quarry (whitish spots of soil) (Photographs by: E. Koptseva)

Fig. 3.3c The second stage of natural recovery defined as a community controlled by the environment (Photograph by: E. Koptseva)

Fig. 3.4 Plant communities in well-drained sites (**a**) and in wet bottom depressions after 40 years of natural recovery (**b**) (Photographs by: E. Koptseva)

structure of recovering plant communities begin to emerge more clearly. However, differences between naturally recovering communities from their natural, untouched habitants still exist; for example the lack of balance between the projective cover of dominant plants. On this stage, for example, communities with young woody plants, shrubs, mosses and lichens already occupy well-drained sites (Fig. 3.4a).

Among the plant groups, erosion tolerant plants often prevail in plant cover, for example dwarf shrub *Empetrum hermaphroditum* Hagerup and lichen *Cladonia gracilis* (L.) Willd. Communities of this stage are only described in "old" quarries which were decommissioned more than 40 years ago.

Final recovery stage is associated with transformation to the original condition (as was before the disturbance) and which is basically not observed even in old quarries because restoration of plant production and resource potential requires for a longer time and is still ongoing. For forest-tundra communities the time to reach the final recovery depends on regeneration of tree layer, with the dominance of native coniferous tree species. After 40 years of natural recovery, only the vegetation of a damp floor of the quarry has returned to resemble closest to the undisturbed fens (Fig. 3.4b).

Finally, during natural recovery, quarry area looks like a mosaic of biotopes, each of which is occupied by the most appropriate type of vegetation. In our investigation we observed that dry young pine forests with lichens can regenerate only in well-drained tops, sedge and cottongrass fens in stagnant water sites, and willows – in flooding positons. Zonal features of vegetation can be observed in recovering plant communities on the steep slopes. Young larch and spruce trees as well as some shrubs (for example cranberries) can settle on the sloping positions as in Fig. 3.3c.

Key features of the natural recovery such as species composition of vascular plants and cryptogams, as well as the ratio of main plant morphology groups (woody plants, forbs, graminoids) change in time. In quarries, the diversity of plant species and their morphology is significantly different from undisturbed zonal habitats especially at the beginning of primary succession. Usually herbs predominate in vegetation of quarries during the first five-ten years, but proportion of woody plants is substantially reduced. In time the ratio between the different types of plants in vegetation cover changes. The proportion of woody plants progressively increases, while part of forbs and graminoids instead decreases. However even in "old" quarries (after 40 years) the ratio between the different types of plants differs from zonal undisturbed communities (Fig. 3.5a).

Increase in the number of plant species is usually observed in the initial stages of primary succession. This process is accompanied by the increasing of number of different types of plant communities. The maximum number of species and the biggest variety of plant communities are observed after 9–13 years from the beginning of natural recovery. After that, the diversity of plant species and their composition decrease due to environmental selection (Fig. 3.5b).

3.2.2   Artificial Restoration of Vegetation

It was shown that vegetation differs from natural one even in "old" quarries that have recovered more than 40 years. Despite the positive successional trends, natural recovery is difficult in the extreme climatic conditions of the Siberian North. Artificial restoration is a mandatory requirement of Russian environmental law. Usually it consists of two phases: engineering and biological restoration.

Before the biological phase of the restoration, the surface of the quarry needs to be levelled out. The reduction of the inclination of the quarry makes the slopes more resistant to erosion (max. 20° slope). This is called as the engineering phase of restoration. The biological stage is to apply peat, fertilizer and plants.

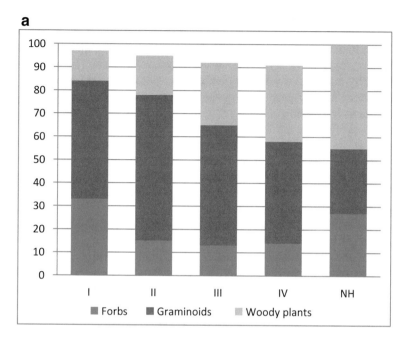

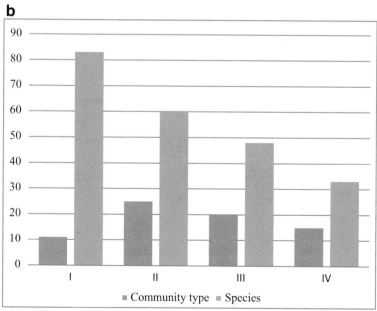

Fig. 3.5 Transformation of basic community parameters in primary succession: (**a**) – the proportion of the different types of vascular plants in the quarries and in natural habitats (NH); (**b**) – the number of plant community types and a number of species. Recovery stages: *I* ecological chaos, *II* the second stage, *III* the third stage; *IV* the final stage

Taking into account the topography of disturbed sites, using a specifically prepared fertilizer mixture and high quality seeds and seedlings is the only effective way of restoration. For example, near Nadym, vegetation of the quarry was restored by pine cultures (*Pinus sylvestris* L. subsp. *lapponica* (Fries ex Hartm.) Hartm. ex Holmb.). Dry and semi-dry young pine forests with moss and lichen cover were formed already after 15 years from the beginning of the restoration (Fig. 3.6a).

Unfortunately, planting of forest cultures is used quite rarely in the Yamal-Nenets district. Most often, the artificial restoration in the region is focused on the sowing of perennial grasses and legumes. This is so-called stimulation of natural regeneration. At the same time, it should be remembered that too dense and high grass cover can prevent the germination of native species including forest shrubs, mosses and lichens. In such case, communities with young pine trees and dense grass cover are developed after 15 years (Fig. 3.6b).

Artificial restoration makes the recovery more uniform but it reduces the diversity of biotopes available for formation of different vegetation types. In addition, the majority of the quarry land-area remains free from vegetation for a long time because a strong northern wind blows the seeds away from the quarry. The presence of pits and shallow depressions instead of completely even surface would contribute to a more active process of overgrow by willow and birch.

At times, attempts to use willow cuttings for slope fixation are being made. Willow species are not demanding on the environment, they have a high vegetative reproduction and they most importantly grow in abundance in surrounding natural vegetation along the riverbanks. As natural recovery demonstrated, in spite of high ecological tolerance willow cuttings take root most successfully only in the bottom of the quarry or in lower part of slopes in wet conditions, but not in long stagnant water. Flat and dry surfaces are not suitable for reforestation by willow cuttings (Fig. 3.7a). Similarly, stagnant water sites are unfavorable for the planted young pine trees (Fig. 3.7b). In both of these cases the natural changes of communities by more adapted to existing environmental factors has occurred.

3.3 Conclusions

This research showed that intensive methods of restoration are required in northern regions of Russia in order to the speedy return of the disturbed lands to any further use. As shown, the natural recovery continues for a long time, spanning over many decades. Effective restoration can help to avoid the ecological chaos stage, makes the succession of vegetation more manageable and finally intensifies the recovery. However, in each case of restoration a thorough preliminary analysis of existing conditions in these disturbed sites is necessary for choosing the most appropriate methods for the restoration. Underestimation of the effect of environmental conditions may considerably reduce the restoration effectiveness and even lead to financial losses if the restoration fails.

Altogether, a combination of intensive artificial restoration in sites with active surface erosion and promotion of the natural regeneration in the most suitable land-

Fig. 3.6 Restoration results: effective pine cultures (*Pinus sylvestris* L. subsp. *lapponica* (Fries ex Hartm.) Hartm. ex Holmb.) in the quarry near Nadym City in 2014 (**a**); pine self-seeding in dense artificial grassland (**b**) (Photographs by: A. Egorov and E. Koptseva)

form positions are the most effective methods for restoration. Unfortunately, in the Yamal-Nenets Autonomous District, artificial recovery is limited due to the lack of special forest nurseries and adapted grass mixtures.

Fig. 3.7 Negative restoration results. Death of willow cuttings and pine tree self-seeding in dry sites (**a**); Disappearing pine cultures and willow self–seeding in stagnant water sites (**b**) (Photographs by: E. Koptseva)

Our previous observations (Egorov 2015) and the other investigations have resulted in the following recommendations for successful restoration of quarries in the forest-tundra and forest zones of the Yamal-Nenets Autonomous District:

- Add peat or forest fertile soil in sufficient quantity.
- Take into account the possibility of erosion even on plots with a slight slope angle.
- Plant annual and perennial grasses; use of rapid growth annual grasses which create protection for perennial species to stabilize soil surface.
- Use of adapted adventive and native plant species.
- Use of leguminous plant species for increasing soil fertility.
- Ensure sufficient tree density during planting.

Acknowledgements This work was supported by the Department of Science and Innovation of the Yamalo-Nenets Autonomous District (grant № 01-15/4, 25.07.2012).

References

Abakumov EV, Maksimova EI, Lagoda AV, Koptseva E (2011) Soil formation in the quarries for limestone and clay production in the Ukhta region. Eurasian Soil Science. T. 44. № 4. C. 380–385

Barrington M (1927) Cajander's Theory of forest types. Ecology 8(1):135–137

Borgergard S-O (1990) Vegetation development in abandoned gravel pits: effects of surrounding vegetation, substrate, and region. J Veg Sci 1:675–682

de Witte LC, Stöcklin J (2010) Longevity in clonal plants: why it matters and how to measure it. A review. Ann Bot 106:859–870

Egorov AA (2015) Features recultivation quarries in the forest-tundra and forest zone of north of Western Siberia. Natural resources and integrated development of coastal areas in the Arctic zone: materials of International scientific conference (29.09-01.10 2015, Arkhangelsk) Arkhangelsk. P. 112-116. (in Rus)

Forbes BC, Ebersole JJ, Strandberg B (2001) Anthropogenic disturbance and patch dynamics in circumpolar Arctic ecosystems. Conserv Biol 15(4):954–969

Soil Sampling and Methods of Analysis (2008) Soil sampling and methods of analysis, Second Edition. Edited by M.R. Carter and E.G. Gregorich. Canadian Society of Soil Science. 1224 p

Sumina OI (1998) The taxonomic diversity of quarry vegetation in North-West Siberia and Chukotka. Polar Geogr 22(1):17–55

Walker LR, del Moral R (2008) Lessons from primary succession for restoration of severely damaged habitats. Appl Veg Sci 12:55–67

Chapter 4
The Concept of Hierarchical Structure of Large Marine Ecosystems in the Zoning of Russian Arctic Shelf Seas

Kirill M. Petrov and Andrey A. Bobkov

Abstract Main features of biogeographical regionalization were developed in previous notes of authors. In this chapter a review on the new information on large marine ecosystems is given based on strong theoretical and empirical material including own research. The originality of method of a research consists that at the description of the sea basin (ecoregion) the three-rank system of units considering zonal, vertical and azonal distinctions of the environment which influence the distribution of marine inhabitants is used. The principles of the regionalization of hierarchical structure are discussed on the example of the Barents Sea.

4.1 Introduction

The system of the Global Ocean bioregionalization developed by Spalding et al. (2007) includes 12 realms, 62 provinces and 232 ecoregions (large marine systems). In the Arctic realm whole marine basins (seas) are treated as ecoregions. Their biogeographical features are determined by the degree of their isolation, surface and deep water circulation system, river runoff, tides, thermohaline regime, waving condition of formation and dynamics of ice cover, bottom relief and geological features, ground deposits, types of shore and coast line configuration. The essential role in the formation of oceanographic regime of water area is coupling the interaction between the sea and atmosphere. As a solution to detailed biogeographical division of large marine ecosystems, a landscape and bionomic approach was suggested (Petrov 2004, 2012), where the patterns of marine organisms distribution are related

K.M. Petrov (✉) • A.A. Bobkov
Institute of Earth Science, Saint-Petersburg State University,
33/35, 10th line, Vasilievsky Island, Saint-Petersburg 199178, Russia
e-mail: k.petrov@spbu.ru; abbk-437@yandex.ru

© The Author(s) 2017
K. Latola, H. Savela (eds.), *The Interconnected Arctic — UArctic Congress 2016*,
Springer Polar Sciences, DOI 10.1007/978-3-319-57532-2_4

to the conditions of their habitat. According to this approach sea bottom organisms which are usually called benthos are consistently found in the combination of relief and types of sea sediments with a specific set of hydrological and hydrochemical factors forming in the whole the large marine ecosystem consisting of local ecoregions. A detailed regionalization of such large marine ecosystems in range of sea basins is necessary for carrying out monitoring, rational use and protection of marine biological resources at different hierarchical levels. Experience of authors of present paper on biogeographical regionalization of the sea basins is based on rich theoretical and empirical material, including on own researches executed earlier summarized in works of Petrov (2009, 2012), Bobkov and Petrov (2013).

4.2 Methodical Approach

The biogeographical heterogeneity of the Global Ocean can be treated from landscape and bionomic positions. This approach is based on statement that the specificity of biotic composition is determined by natural conditions. The formation of a bionomic structure in a marine ecoregion is influenced by oceanographic processes which cover water masses, the bottom and boundary layers of the atmosphere (Petrov 2004, Spalding et al. 2007). The landscape-bionomic differentiation of large marine ecosystems reflects three directions of these processes: zonal (latitudinal), vertical (deep) and azonal (geomorphological). In a multicomponent scheme of interactions, two main functional links are distinguished: hydrological and geologo-geomorphological (Petrov 2004). Water masses serve as the main ecological factor of biogeographical regionalization. Warm Atlantic waters are delivered into the Barents Sea by Norwegian, West Spitsbergen, Nordcape and West Greenland currents. Cold waters and ice are taken away from the Arctic with East Greenland and Labrador currents. The influence of a hydrological factor is manifested in zonal differentiation of the sea basins. The geological structure and bottom relief constitute a lithogenic basis on which underwater landscapes are formed. Relief and

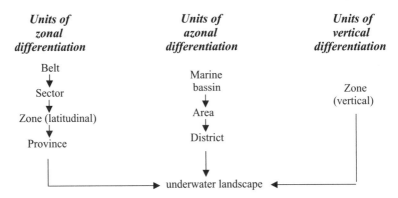

Fig. 4.1 Three-rank system of units of landscape-bionomic regionalization (Petrov 2004)

types of ground define the geomorphological subdivision of the bottom which can be used as the basis for studying benthic organisms distributions. The geologo-geomorphological structure of the sea floor serves as a framework for defining a system of azonal units regionalization which are not connected with system of zonal units. The both above mentioned feedback loops bear certain environmental pressure affecting to third connection – the bionomic one. Thus, a three-rank system of units at regional level was earlier proposed by Petrov (2004) the essence of which is given in Fig. 4.1. It can be seen that the underwater landscape connects all three units of differentiation. It is characterized by a common geological structure, relief and homogeneous hydroclimate with a corresponding set of bottom organisms and is accepted as an initial unit in hierarchical structure of marine ecoregion.

4.3 Units of Zonal Differentiation

The Arctic Ocean and its seas belong to the Arctic geographical belt subdivided into Polar and Arctic nature zones. In the Arctic zone change of environmental conditions occurs in both latitudinal and longitudinal (sectors) directions. Four climatic sectors may be distinguished in the North Polar area: Atlantic, Siberian, Canadian and Pacific (Korotkevich 1985). The Barents Sea belongs to the Atlantic coastal sector.

The division of the Barents Sea Arctic zone into arctic and subarctic provinces is determined by the influence of warm and cold currents forming a stable water mass transfer with characteristic zonal groups of plankton and benthos (Figs. 4.2 and 4.3).

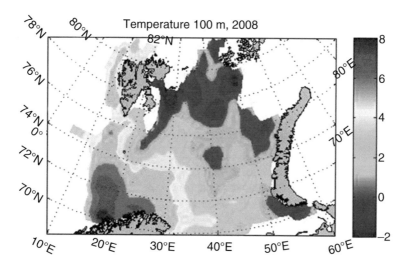

Fig. 4.2 Average water temperature distribution at 100 m depth in summer of 2008 illustrating spreading of warm and cold waters (Skern-Mauritzen and Fall 2010). *Yellow-brown color* of spectrum corresponds to the boreal zone, *turquoise-green* – to the subarctic province, and *blue* – to the arctic province of arctic zone

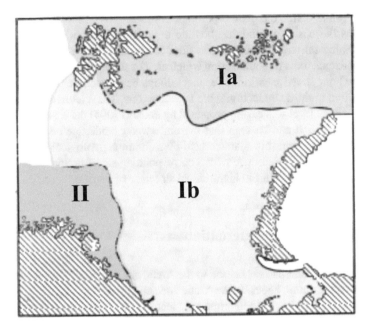

Fig. 4.3 Zonal differentiation of the Barents Sea: *I* arctic zone: *Ia* arctic province, *Ib* subarctic province, *II* boreal zone, extrazonal Barents Sea province (Modified from Zenkevitch 1963)

The central part of the basin is situated in the subarctic province. The south-western part of the sea is the most warm area heated by Atlantic waters and belongs to boreal zone. The Barents Sea extrazonal province (Petrov 2009). Located at the intersection of warm and cold waters, the Barents Sea contains a biota which constantly experiences changes with either the onset of warm-water forms or their replacement with cold-water forms. Therefore, the zonal boundaries are vague and encapsulate the arctic and boreal zones through narrow dynamic transitions.

4.4 Units of Vertical Differentiation

The vertical differentiation reflects the change of environmental conditions with depth. The properties of vertical zones depend on the natural zone (province) in which they are formed: boreal, subarctic or arctic. The main units of the vertical differentiation are littoral, sublittoral and elitoral zones which are subdivided into floors and stages. An example from the East Murman (the Kola Peninsula) is given. The <u>littoral</u> zone (with the tide amplitude of about 3 m) is divided into two floors. An indicator of the first floor is the belt of *Fucus vesiculesus*; the second one seen in syzygy tides is characterized by a *Laminaria digitata*. The <u>sublittoral</u> zone is divided into three floors. Its components are shown in Fig. 4.4. The upper floor consists of two stages. In the first (*I a*) the community of *Laminaria digitata*

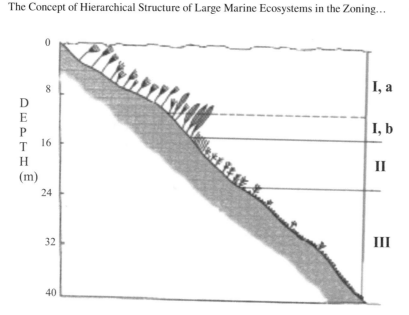

Fig. 4.4 Vertical differentiation of sublittoral zone of the underwater coastal slope of East Murman: *I–III* – sublittoral floors: *I* upper floor, *Ia* upper stage (community of *Laminaria digitata*), *Ib* lower stage (community of *Alaria esculenta*), *II* middle floor (community of *Chorda tomentosa* + *Odonthalia dentata*), *III* ground floor (community of *Balanus balanus*) (Modified from Propp 1971)

prevails. In the second *(Ib)* – community of Alaria esculenta. In the middle floor *(II)* the community of *Chorda tomentosa* + *Odonthalia dentata* dominates; in the ground floor – community of *Balanus balanus* (Propp 1971). The upper (surface) stage is located on the border between the atmosphere and sea surface, it is subject to the seasonal fluctuations depending mainly on the budget of solar radiation. On the bottom of the Barents Sea the thermal regime is defined distribution of arctic, subarctic and boreal water masses. In underwater landscapes, where the temperature near the sea floor is below freezing all year round, high arctic fauna dominates. A diver moving from the littoral to the sublittoral zone will find changes in fauna distribution from boreal to high arctic one, even in the same area that confirms its dependence on the water temperature.

4.5 Units of Azonal Differentiation

The azonal features of marine basin are presented to us as following: the size, depth and bottom hollow form, a relief of the shore and degree of isolation from the Global Ocean. In general, the distribution of life forms on the seabed is controlled by properties of geological structure, bottom relief and sea sediments and these items gives a reason to distinguish a system of azonal units shown in Fig. 4.1, namely: the marine basin, area, district, region (landscape).

4.6 Discussion

Landscape-bionomic concept characteristic of the Barents Sea as of the large marine system marked in scheme of ecoregions of Spalding et al. (2007) shown in Fig. 4.5 under # 18 was elaborated. Its structural elements became reason to discuss. The concept is based on the WWWF's Arctic Programme (*Barents Sea ....* 2003) visualized on a map in Fig. 4.6. The map was taken as a basis issue where ecoregions shown in Fig. 4.6, are correlated with features deriving from joint analysis of distributions of principal geomorphological elements and macrobenthos shown in Figs. 4.7 and 4.8 respectively. A brief description of these ecoregions is provided below.

1. *Southwestern ecoregion* in Fig. 4.6 (1 a, Norway near-shore areas, and 1b, the Kola Peninsula near-shore areas) corresponds to Southwestern geomorphological area in Fig. 4.7. The coastline with narrow shelf is washed by waters of the North Atlantic current. The boreal fauna with relic forms in deep fjords prevails. Owing to the influence of warm Atlantic water, coastal sites have the greatest variety of benthos presented with brown and red algae, sessile and vagile sestonophagous – filtrators on a stony slope. The variety and productivity of benthos decreases with depth.
2. *Ecoregion of the Pechora Sea* in Fig. 4.6 corresponds to Kanin-Pechora flat geomorphological area in Fig. 4.7. In shallow waters on sandy-mud grounds sessile and vagile organisms prevail: sestonophagous, detritivores and ground feeders inhabiting the seabed. The dominating communities are bivalve molluscs *Ciliatocardium ciliatum, Macoma calcarea* and *Serripes groenlandicus*.
3. *Ecoregion of the Central basin south from the polar front* in Fig. 4.6 corresponds to the southern part of the Central Barents rift in and the West Barents Sea tectonic and geomorphological areas in Fig. 4.7 and represents a wide transition zone between the Atlantic and Arctic waters. Deep-water communities of ground

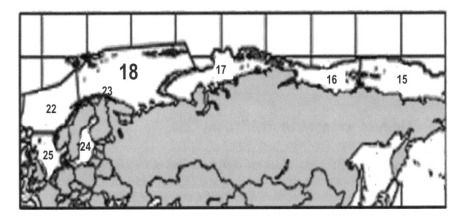

Fig. 4.5 Ecoregions of Euro-Asian shelf of Arctic realm modified from Spalding et al. (2007). Fragment. 18 – North and East Barents Sea ecoregion

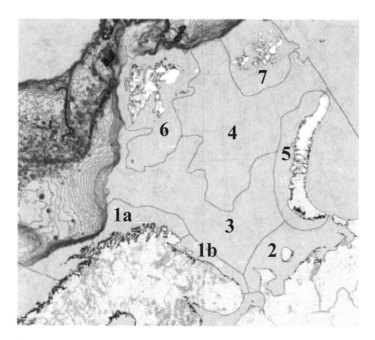

Fig. 4.6 Ecoregions of the Barents Sea: – South-Western (*1a* near-shore areas of Norway, and *1b* the Kola Peninsula near-shore areas), *2* Pechora Sea, *3* Central basin south from the polar front, *4* Central basin north from the polar front, *5* Novaya Zemlya shore, *6* Svalbard Archipelago and banks of Spitsbergen, *7* Franz Josef Land Archipelago (Modified from Barents Sea Ecoregion 2003)

feeders polychaeta and sipunculida (*Golfingia* sp., *Spiochaetopterus typicus, Ctenodiscus crispatus*) as well as the community of holothurian (*Trochostoma* sp.) are characteristic. The Eastern part of this ecoregion coincides with the Central Lowland geomorphological area inhabited by boreal-arctic and arctic fauna of detritivores. Community of bivalve molluscs can be observed of family Astartidae (*Elliptica elliptica* and *Astarte crenata*), and deep-water community of polychaeta, sipunculida and holothurian can be found.

4. *Ecoregion of the Central basin north from the polar front* in Fig. 4.6 is situated in sphere of spreading of Arctic water masses. It occupies the northern part of the Central Barents Sea rift and belongs to the North Barents Sea geomorphological area in Fig. 4.7. This is a zone of detritivores collecting detritus from a seabed. The deep-water community includes *Ophiopleura borealis* and foraminifer *Hormosina globulifera*, and also the community of bivalve molluscs of family Astartidae (*Elliptica elliptica* and *Astarte crenata*).

5. *Ecoregion of the Novaya Zemlya shore* in Fig. 4.6 corresponds to the Novaya Zemlya tectonic and geomorphological areas in Fig. 4.7. It is washed by the Arctic water mass with some influence of Atlantic water in the west. On sandy-mud bottom grounds the trophic groups comprise sessile and vagile sestonophagous-filtrators *Hyatella arctica*, *Strongylocentrotus* sp., whereas *Ophiura robusta, Balanus balanus* are dominating in bentic community.

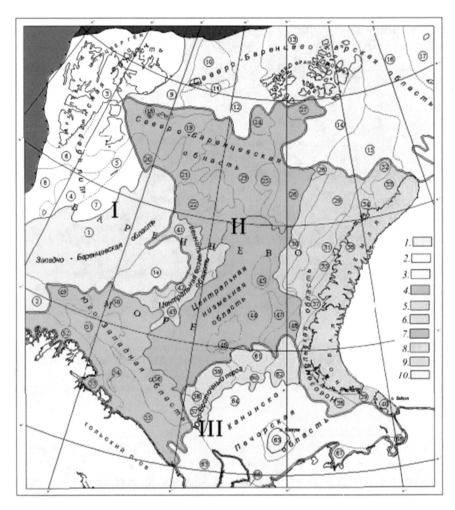

Fig. 4.7 Geomorphologic regionalization of the Barents Sea after *Zinchenko* (2000). **I–III provinces,** *1–10* **areas,** *1–68* local geomorphological elements of less scale. **I Outer shelf province,** *areas*: *1* West Barents Area of trenches and hollows (items 1–2), *2* Spitsbergen Upland Area (items: 3–8), *3* North Barents-Kara Area of elevations (items: 9–16). **II: Inner shelf province,** *areas*: *4* North Barents Plain Area (items: 17–27), *5* Novaya Zemlya Area of linear uplands and trenches (items: 28–40), *6* Central Upland Area (items: 41–43), *7* Central Lowland Area (items: 44–48), *8* South-Western Area mainly of linear uplands and trenches (items: 49–56). **III: Shallow-water province,** *areas*: *9* Kanin-Pechora's Area, South-Eastern Rapid (items: 57–62), *10* Kanin-Pechora's Plain Area (items: 63–68)

6. *Ecoregion of the Svalbard Archipelago and the in banks of Spitsbergen* in Fig. 4.6 is located inside the Spitsbergen highland geomorphological area in Fig. 4.7 filled by Arctic water mass but influenced by the North Atlantic. It is characterized by high biodiversity. Trophic groups include macrophytobenthos, sessile and vagile sestonophagous-filtrators. On stony bottom grounds, the community of brown algae dominates. The community of sessile sestonophagous is formed

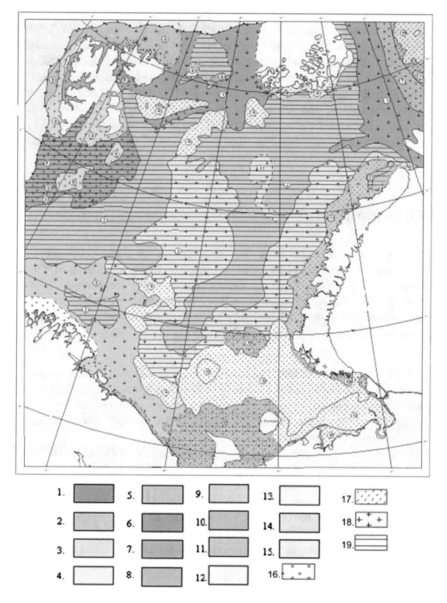

Fig. 4.8 Distribution of macrobenthos in the Barents Sea and adjacent waters (Kiyko et al. 2006). Communities: *1 Ophiopleura borealis + Hormosina globulifera*, *2 Polychaeta + Sipunculoidea* (*Golfingia sp.*), *3* Trochostoma sp., *4 Elliptica elliptica + Astarte crenata*, *5 Brisaster fragilis*, *6* soft bottom community adjacent to Svalbard, *7* community of Saint Anna trench slopes, *8 Strongylocentrotus sp. + Ophiopholis aculeata*, *9* shoal community of sessile filter-feeders adjacent to Svalbard, *10* shoal community of sessile filter-feeders on *Lithothamnion sp.*, *11* shoal community adjacent to western coast of Novaya Zemlya and Vise Island, *12* bivalve mollusc *Tridonta borealis*, *13* bivalve molluscs *Ciliatocardium ciliatum + Macoma calcarea + Serripes groenlandicus*, *14* bivalvie molluscs of Ushakov Plateau, *15* bivalve mollusc *Macoma fusca*; *16–19* regions of accumulation of pollutants: *16* chlorineorganics, *17* Fe and Mg, *18* polymetallics (Zn, Sn, Cu, Ni), *19* all pollutants by ground feeders

by *Balanus balanus, Tridonta borealis, Hydroidea* var.*, Nephthys* sp. On soft grounds, *Spiochaetopterus typicus, Nephthys* sp., *Lumbrinereis fragilis* prevail.

7. *Ecoregion of the Franz Josef Land A*rchipelago in Fig. 4.6 occupies the North Barents-Kara tectonic and geomorphological areas in Fig. 4.7 and is washed by the Arctic waters. Despite of its northern location, the coastal zone is characterized by both high biodiversity and productivity. The group of inhabitants includes acrophytobenthos, and sessile and vagile sestonophagous-filtrators. Among ground feeders are the communities of polychaeta and sipunculida, as well as the community of holothurian. The slope community of the Saint Anna trench is formed by *Ascidiacea* var., *Nephasoma minuta, Thenea muricata.*

4.7 Conclusions

In this article the current literature and knowledge on the hierarchical structure of large marine ecosystems was reviewed and based on that it was presented how the Russian Arctic shelf seas fit to the system. The emphasis has been done on the Barents Sea. The following conclusions based on experience of authors on biogeographical regionalization of sea basins including their rich theoretical (bibliographical) sources and own empirical material executed earlier issues, are summarized below.

1. The principles of the hierarchical zoning system of units were discussed by using the large marine ecosystem of the Barents Sea as an example.
2. Landscape-bionomic differentiation reflects three components of the oceanographic process: zonal, vertical and azonal., and three-rank system of units at regional level was introduced.
3. The formation of the marine basin bionomic structure is controlled by the oceanographic processes. In a multicomponent scheme of interactions the following functional links are distinguished: hydrological and geologo-geomorphological ones. Both of them cause a certain environmental pressure, effecting the structure and functions of the third link – the bionomic one.
4. The regionalization unit system is essential for monitoring, rational use and protection of marine biological resources at different hierarchical levels.

References

Barents Sea Ecoregion (2003) A biodiversity assessment. Proceedings of WWF biodiversity workshop. (http://assets.wwf.no/downloads/wwfrapport_barentsecoregionassessment_nov2003.pdf)
Bobkov AA, Petrov KM (2013) The Sea of Okhotsk as the large marine ecosystem. Part 1: methodology. In: Proceedings of the 28th international symposium on Okhotsk Sea & Sea Ice (OSCORA). Mombetsu, Hokkaido, Japan, Feb 17–21, 2013, pp 324–327

Kiyko OA, Kulakov MYu, Timofeyev SF, Chernova NV (2006) Ecosystem of the Barents and Kara seas coastal segment. The Global Coastal Ocean. Interdisciplinary Regional Studies and Syntheses. In: Robinson AR, Brink KH (eds) The sea: ideas and observation on progress in study of the seas. Vol. 14. Part B. Harvard University Press, Cambridge, MA, pp 1139–1176

Korotkevich ES (1985). Physico-geographical regionalization. Geohraphy of the World Ocean. North Polar and Southern oceans. In: Troshnikov AF, Sal'nikov SS (eds) Nauka, Leningrad, pp 120–129 (in Russian)

Petrov KM (2004) Bionomy of ocean. SPb.: SPb Press, 242 p (in Russian)

Petrov KM (2009) Large marine ecosystems: construction principles for a hierarchical system of arctic seas regionalization exampled with Barents Sea. J Biosphere 1(2):133–152. (in Russian)

Petrov KM (2012) Theory of hierarchical organization of large marine system. Palmarium Academic Publishing, Saarbrücken. 260 p. (in Russian)

Propp MW (1971) Ecology of coastal bottom communities of Murman Shore of Barents Sea. Nauka, Leningrad. 128 p. (in Russian)

Skern-Mautitzen M, Fall J (2010) Dolphin densities and distributions in the Barents Sea and potential influences by the recent temperature increase. 13 p. (http://www.iwcoffice.org/_documents/sci_com/workshops/SmandCC/SC-N10-CC5.pdf)

Spalding MD, Fox HE, Allen GR, Nick Davidson N, Ferdana ZA, Finlayson M, Halpern BS, Jorge MA, Lombana A, Lourie SA, Martin KD, MCManus E, Molnar J, Recchia CA, Robertson J (2007) Marine ecoregions of the world: a bioregionalization of coastal and shelf areas. J Biosci. 57(2): 573–583. (http://www.biosciencemag.org)

Zenkevitch LA (1963) Biology of the seas of the USSR. Wiley-Interscience, New York. 955 p

Zinchenko AG (2000) Map of geomorphologic regionalization of Barents Sea. Atlas of Barents Sea. Funds of VNIIOkeangeologia, Saint-Petersburg. Computer variant of the map (in Russian)

Chapter 5
Changing Climate and Outbreaks of Forest Pest Insects in a Cold Northern Country, Finland

Seppo Neuvonen and Heli Viiri

Abstract Pest insect population dynamics are species specific and complex due to nonlinearities and interactions among different trophic levels. Consequently, the impacts of climate change on pests are also species specific and they are often difficult to predict. However, there are some clear examples of increasing forest pest risks due to a warming climate. The damage caused by the Eurasian spruce bark beetle has recently increased in Finland as a consequence of more frequent storm damage and longer growing seasons. In a warming climate, timely salvage and sanitation cuttings will be needed to guarantee the sustainability of the forestry. Several defoliating pests overwinter in the egg stage. Warmer winters may not kill the eggs and, therefore, the incidence of outbreaks is predicted to increase in the northern and continental areas. The most important societal implications will be due to Geometrids attacking subarctic mountain birch forests. Together with heavy reindeer grazing, Geometrids reduce the resilience of the ecosystem and they are threatening the sustainability of local livelihoods.

5.1 Introduction

In the Boreal zone, insects have had an essential role in the succession dynamics and in starting the succession process again in natural forests. However, in northern Europe the forests have been under intensive forestry for a long time. This has fragmented the landscape structure (Kouki et al. 2001), so that large and extensive insect outbreaks have occurred only rarely. The effects of changing climate on the population dynamics of insect pests are complex and species specific, and only some species are projected to increase in a warming climate (Björkman and Niemelä 2015).

Major pest insects are rare most of the time but they cause damage every now and then. For example, the plot level probability of pine sawfly outbreaks occurring at

S. Neuvonen (✉) • H. Viiri
Natural Resources Institute, Joensuu, Finland
e-mail: seppo.neuvonen@luke.fi

© The Author(s) 2017
K. Latola, H. Savela (eds.), *The Interconnected Arctic — UArctic Congress 2016*,
Springer Polar Sciences, DOI 10.1007/978-3-319-57532-2_5

least once in 20 years varied from about 10% in the most fertile site types to 30–40% in sub-xeric to xeric pine stands (Nevalainen et al. 2015).

Insect outbreaks can be classified into different types based on the population dynamics involved (Berryman et al. 1987):

(A) Regularly cyclic outbreaks: in northern Europe the Geometrid moths (*Epirrita autumnata*, *Operophtera brumata*) that defoliate mountain birches belong to this type (Haukioja et al. 1988).
(B) Eruptive outbreaks occur at irregular intervals. They may be triggered by specific environmental conditions (e.g. drought) and individual outbreaks are short. Pine sawflies (*Neodiprion sertifer*, *Diprion pini*) exemplify this type (Hanski 1987; Juutinen 1967). Another pest that has eruptive outbreaks is the European spruce bark beetle (*Ips typographus*), which is mainly regulated by resource availability (Økland and Bjørnstad 2006).
(C) Sustained outbreaks that may last for several years also occur at irregular intervals and they are triggered by environmental conditions. Although these kinds of outbreaks are rare in northern Europe, a recent example can be found in the damage caused by the large web-spinning sawfly (*Acantholyda posticalis*) in Finland and in Estonia (Pouttu and Silver 2016; Voolma et al. 2009).

A common feature to all outbreak types is that there is a difference of several orders of magnitude in insect densities during the low density (endemic) phase versus the outbreak (epidemic) phase (Berryman et al. 1987; Hanski 1987). Thus, the increase phase from endemic to epidemic densities requires normally at least 2 to 4 years, during which the environmental conditions must remain suitable for rapid population growth.

The climatic conditions in the boreal zone of northern Europe show high year to year variability. This variability can interact with other factors affecting the population dynamics of forest pest insects (Neuvonen and Virtanen 2015). Recent patterns in the outbreaks of the forest pests in relation to recent climatic changes are reviewed and discussed in this chapter.

5.2 The Life Cycles of Pest Insects in Relation to Recent Climate Change in Finland

Pest insects have complex life cycles and life stages. They live in a variety of microhabitats that experiencing very different climatic conditions. Consequently, considering the variable responses of different life stages to changing climate is essential to understand the impacts of climate change on pests (Kingsolver et al. 2011). Many of the forest defoliators in the Boreal zone overwinter in the egg stage and feed on early season foliage (Hunter 1991). The climatological winters (temperatures <0 °C) last several months in northern areas, experiencing occasionally extreme temperatures that are much lower than seasonal averages. These factors have important consequences for insect pests and for other ecological processes (Neuvonen et al. 1999; Williams et al. 2014).

Extremely cold temperatures can kill eggs that are overwintering in the canopy (Austarå 1971; Nilssen and Tenow 1990). Consequently, higher winter minimum temperatures will increase the outbreak risks of pest species overwintering as eggs (Virtanen et al. 1996; Virtanen et al. 1998), but may not affect pests overwintering in the soil. The latter are normally protected by insulating snow cover and are not so sensitive to variations in air temperatures (Virtanen and Neuvonen 1999a).

The annual mean temperature in Finland has risen by a total of 2.3 °C from the mid-nineteenth century to the present (i.e. 0.14 °C per decade) (Mikkonen et al. 2015). The largest warming has been observed in winter temperatures. The spring (March–May) has also warmed more than the annual average, but during the summer months (i.e. during the time when most forest pests are actively feeding) there has been only very little or no warming (Mikkonen et al. 2015). The warming has not been even. For example, between the 1940s and the 1960s the climate did not warm (Mikkonen et al. 2015), but from the end of the 1960s onwards the mean daily temperatures have warmed on average by 0.3 °C per decade (Aalto et al. 2016).

The average temperature sums have increased by about 20% during the past 20 years and the incidence of storm damage has also increased during the last decade. This has increased the risk of spruce bark beetle damage, especially in southern Finland (Viiri and Neuvonen 2016). Given that the temperature during the summer months (June–August) has not increased much (Mikkonen et al. 2015; Neuvonen and Viiri 2015), the increase in temperature sums is mainly due to increases in spring and autumn temperatures.

When evaluating the potential impacts of climate change on forest pest insects, the following should be kept in mind:

(1) Forest insect pests do not generally experience the weather and climatic conditions recorded at weather stations. The effects of microclimates should be considered when estimating the ecological impacts of climate change (Daly et al. 2010; Potter et al. 2013). GIS techniques can be used when estimating the values of target variables between or around the weather stations (Virtanen et al. 1998).

(2) The inter-annual variability of temperatures in northern areas is very large. For example, in Finland the range of variation in monthly mean temperatures within a decade is 10–15 °C during winter and about 5 °C during summer (Neuvonen and Viiri 2015). This variation is about an order of magnitude larger than the observed or projected decadal trends in temperatures.

5.3   Birch Defoliators

Cyclic outbreaks (8–11 year intervals) of defoliating Geometrids (*E. autumnata, O. brumata*) are typical for the mountain birch forests of north western Europe (Babst et al. 2010; Tenow 1972). In Finnish Lapland, the climate is more continental than in northern Sweden and Norway (Neuvonen et al. 2005), and low winter temperatures have historically reduced the regularity of outbreaks (Neuvonen et al. 1999).

Fig. 5.1 Autumnal moth larvae have defoliated mountain birch forest in northern Sweden, affecting ecosystem services, local livelihoods, and the touristic value of the landscape (Photograph by Seppo Neuvonen)

The intensity of the peaks has varied considerably. The largest outbreaks have killed hundreds of square kilometres of birch forest (Seppälä and Rastas 1980). They can have devastating effects on ecosystem services and the condition of reindeer pastures (Biuw et al. 2014; Jepsen et al. 2013) (Fig. 5.1). Due to warmer winters that are not capable of killing the overwintering eggs, the incidence of outbreaks is predicted to increase in future in the continental areas of northern Europe (Ammunét et al. 2012; Virtanen et al. 1998). The number of defoliation years has increased because outbreaks of these two species have followed each other (Klemola et al. 2008).

The largest and most devastating outbreak was that of the mid-1960s in Utsjoki (the northernmost municipality in Finland), which changed very large areas of mountain birch woodland to secondary Tundra due to low recovery under heavy reindeer grazing pressure (Chapin et al. 2004a; Kallio and Lehtonen 1973). In the mid-1990s, the more southern parts of Finnish Lapland experienced birch defoliation, where old birches were attacked mainly at higher altitudes (Ruohomäki et al. 1997). In 2004–2005, mountain birch forests in Enontekiö (NW Finnish Lapland) experienced heavy defoliation but the birch forests apparently recovered quite well during the subsequent years (Kopisto et al. 2008).

The first recorded outbreak of Winter moth (*O. brumata*) in Finnish Lapland was at the start of this century (Jepsen et al. 2008). It caused serious defoliation of birch in a 400 km² area in Kaldoaivi wilderness area in Utsjoki during 2006–2008 (Jepsen et al. 2009; Santonen 2011). There was no refoliation after this outbreak and dwarf shrubs were also destroyed in the ground layer. This caused extensive changes in ecosystem functions (Biuw et al. 2014) and societal impacts since reindeer pastures were damaged in large areas. This risked the sustainability of local livelihoods (Chapin et al. 2004b; Lempa et al. 2005).

5.4   Pine Defoliators

In northern Europe, the most common defoliating insects on Scots pine are *N. sertifer*, *D. pini* and *Bupalus piniarius*. Outbreaks typically occur in graded sandy soils; that is, on drier and less fertile forest sites (Larsson and Tenow 1984; Nevalainen et al. 2015). Regional *N. sertifer* epidemics have occurred every 10–20 years in southern Finland and this has caused the defoliation of large areas (Juutinen 1967). Although *N. sertifer* damage may look serious, forests typically recover because new shoots remain undamaged. Outbreaks occur at irregular intervals and normally they only last 2–3 years (Hanski 1987; Soubeyrand et al. 2010). The outbreaks end due to a virus disease of the pest and/or increased parasitism (Juutinen 1982; Olofsson 1987). Earlier, it was common to use biological control (Nucleopolyhedrosis virus) against *N. sertifer* (Juutinen 1982), but this virus is no longer allowed to be marketed in the EU.

Females of *N. sertifer* lay eggs into needles, where the eggs overwinter predisposed to low winter temperatures. The eggs can stand −36 °C in mid-winter (Austarå 1971). Colder winter temperatures than this have been common in northern and eastern Finland, which has meant that outbreaks have been rare in these areas (Virtanen et al. 1996). An exceptional case is the *N. sertifer* damage at pine tree line areas in Saariselkä (Finnish Lapland) (Niemelä et al. 1987) where the eggs survive at higher altitudes due to strong temperature inversions in winter.

Normally the common pine sawfly (*D. pini*) causes more local, more irregular and more serious damage because the larvae gnaw all needle classes at the end of summer. If damage continues several years in the same area, then the mortality of trees increase and other pests such as bark beetles attack the trees (Annila et al. 1999). In Finland, the outbreaks of *D. pini* have been less common than those of *N. sertifer*, but during 1997–2000 there was an exceptionally large outbreak of *D. pini* in the central parts of the country (Nevalainen et al. 2010). A rough estimate was that pine forests experienced damage in an area of 500,000 ha, from which 200,000 ha had moderate to heavy damage (Varama and Niemelä 2001). The causes of this outbreak remain unknown.

It has been predicted that the outbreak range of *N. sertifer* will expand in eastern and northern Finland if winter minimum temperatures increase (Virtanen et al. 1996). However, this prediction does not apply to *D. pini*, which overwinters as cocoons in the soil, protected from cold temperatures by the snow cover. Furthermore, clear predictions about the population dynamics of pine sawflies in a changing climate might be impossible because the mortality rates are strongly affected by predation by small mammals (Hanski and Parviainen 1985), and the population dynamics of small mammals are complex and rather unpredictable.

Fig. 5.2 Spruce bark
beetles have first
reproduced in storm
damaged spruce trees
(foreground), and during
the following summer they
have attacked and killed
standing spruces
(Photograph by Seppo
Neuvonen)

5.5 Spruce Pests

Spruce bark beetle, *Ips typographus* L, is the most severe pest on Norway spruce in
Eurasia. It has caused remarkable forest damage in many European countries
(Schelhaas et al. 2003). In Finland, spruce bark beetle damage has been at low level
when compared to the other Nordic countries. However, from the year 2010
onwards, outbreaks of *I. typographus* and other bark beetles attacking spruce have
increased in southern Finland. In summer 2010, thunder storms caused damage in
large areas of central and eastern Finland. Parts of damaged trees remained in forest,
which contributed to the growth of the population level (Viiri et al. 2011). Summer
2010 was also hot and dry in large areas of southern Finland, which lowered the
resistance of spruce trees and predisposed them to bark beetle damage (Fig. 5.2).

  Spruce bark beetle successfully breeds in fresh logged Norway spruce timber
and windblown trees (Eriksson et al. 2008). It can attack healthy trees when the
population level is high (Økland and Bjørnstad 2006). The risk of consequential tree
deaths will increase considerably when the amount of windblown trees increases
(Eriksson et al. 2007). Old growth forests, warm forest edges, fresh clear-cut borders and dry sites are especially vulnerable to damage.

The second generation of spruce bark beetle was noticed for the first time in Finland in 2010 (Pouttu and Annila 2010). The development of the second generation remained mainly at larval and pupal stages, which cannot normally survive the winters in Finland (Annila 1969). Even though the bark beetle population size did not grow with new overwintering adults, more damage was caused by extra attacks on living trees. In addition, in southern Sweden there were observations of the development of two generations of spruce bark beetle after the Gudrun storm in 2005 (Långström et al. 2009).

The warming climate has made conditions more favourable to spruce bark beetle in the northern part of Europe (Økland et al. 2015). Longer growth periods and increased temperature sums have enabled the development of more sister broods and even the development of the second generation in some summers (Neuvonen et al. 2016; Wermelinger and Seifert 1999; Öhrn et al. 2014). Shorter periods of frozen ground and thunder-storms in the summertime have increased the amount of dead wood in the forests, which favours breeding bark beetles (Eriksson et al. 2007). Pheromone monitoring started in 2012 and it has shown that population levels have been at epidemic level since 2013 in many locations in southern Finland (Neuvonen et al. 2016).

5.6 Conclusions and Future Prospects

There are several sources of uncertainty when the impacts of climate change on pest insect outbreaks are predicted. First, the different global climate models and alternative emission scenarios produce large variation in predicted climatic outcomes (Jönsson and Bärring 2011; Ruosteenoja et al. 2016). Downscaling to regional and local levels and to microclimates brings more uncertainty to what will happen in the specific microhabitats where the pest insects are living (Neuvonen and Virtanen 2015; Potter et al. 2013).

Other types of uncertainty arise from the complexity on pest insect population dynamics. These are species specific, and include nonlinearities and time-delays, which may lead into chaotic dynamics (May 1976). Further complexities arise from the interactions among different trophic levels and the indirect effects of climate change via natural enemies (Davis et al. 1998; Virtanen and Neuvonen 1999b).

Given the difficulty in predicting climate change and its impact on insect outbreaks, the focus here is only on two systems where pests have the most important societal implications.

Geometrids Attacking Subarctic Mountain Birch Forests When multiple stressors like moth outbreaks and heavy reindeer grazing (Biuw et al. 2014; Tenow et al. 2005) reduce the resilience of the ecosystem, the changes can be drastic and almost irreversible (Chapin et al. 2004b). Reduced reindeer densities and changes in the seasonal patterns of grazing (pasture rotation) will be necessary for better sustainability of reindeer herding (Wielgolaski et al. 2005).

Bark Beetles Attacking Norway Spruce The risk of bark beetle outbreaks will probably remain high in southern Finland in a warming climate (Viiri and Neuvonen 2016). The most efficient way to control spruce bark beetle damage is to remove damaged and attacked trees from forest before new progenies emerge (Stadelmann et al. 2013). In a warming climate, the reduction of spruce bark beetle risks with management actions (timely salvage and sanitation cuttings) is urgently required to guarantee the sustainability of forestry, especially because of the high economic importance of Norway spruce. Continuous monitoring of population levels and phenological surveys are needed for accurate risk estimates and as a basis for timely advice to forest owners about the best management practices (Viiri and Neuvonen 2016). Logging of trees that are windblown at summertime will be more urgent because the swarming time of spruce bark beetles is longer than it used to be (Neuvonen and Viiri 2015; Öhrn et al. 2014).

References

Aalto J, Pirinen P, Jylhä K (2016) New gridded daily climatology of Finland – permutation-based uncertainty estimates and temporal trends in climate. J Geophys Res-Atmos. doi:10.1002/2015JD024651

Ammunét T, Kaukoranta T, Saikkonen K, Repo T, Klemola T (2012) Invading and resident defoliators in a changing climate: cold tolerance and predictions concerning extreme winter cold as a range-limiting factor. Ecol Entomol 37:212–220

Annila E (1969) Influence of temperature upon the development and voltinism of *Ips typographus* L (Coleoptera: Scolytidae). Ann Zool Fenn 6:161–207

Annila E, Långström B, Varama M, Hiukka R, Niemelä P (1999) Susceptibility of defoliated Scots pine to spontaneous and induced attack by *Tomicus piniperda* and *Tomicus minor*. Silva Fenn 33:93–106

Austarå Ø (1971) Cold hardiness in eggs of *Neodiprion sertifer* (Geoffroy) (Hym, Diprionidae) under natural conditions. Norsk entom Tidsskr 18:45–48

Babst F, Esper J, Parlow E (2010) Landsat TM/ETM+ and tree-ring based assessment of spatio-temporal patterns of the autumnal moth (*Epirrita autumnata*) in northernmost Fennoscandia. Remote Sens Environ 114:637–646

Berryman AA, Stenseth NC, Isaev AS (1987) Natural regulation of herbivorous forest insect populations. Oecologia 71:174–184

Biuw M, Jepsen J, Cohen J, Ahonen SH, Tejesvi M, Aikio S, Wäli PR, Vinstad OPL, Markkola A, Niemelä P, Ims RA (2014) Long-term impacts of contrasting management of large ungulates in the Arctic Tundra–Forest ecotone: ecosystem structure and climate feedback. Ecosystems 17:890–905

Björkman C, Niemelä P (eds) (2015) Climate change and insect pests. CAB International, Wallingford

Chapin FS III, Callaghan TV, Bergeron Y, Fukuda M, Johnstone JF, Juday G, Zimov SA (2004a) Global change and the boreal forest: threshold, shifting states or gradual change? Ambio 33:361–365

Chapin FS III, Peterson G, Berkes F (18 authors) (2004b) Resilience and vulnerability of Northern regions to social and environmental change. Ambio 33:344–349

Daly C, Conklin DR, Unsworth, M.H (2010) Local atmospheric decoupling in complex topography alters climate change impacts. Int J Climatol 30:1857–1864

Davis AJ, Jenkinson LS, Lawton JH, Shorrocks B, Wood S (1998) Making mistakes when predicting shifts in species range in response to global warming. Nature 391:783–786

Eriksson M, Neuvonen S, Roininen H (2007) Retention of wind-felled trees and the risk of consequential tree mortality by the European spruce bark beetle *Ips typographus* in Finland. Scand J Forest Res 22:516–523

Eriksson M, Neuvonen S, Roininen H (2008) *Ips typographus* (L.) attack on patches of felled trees: "Wind-felled" vs. cut trees and the risk of subsequent mortality. For Ecol Manag 255:336–1341

Hanski I (1987) Pine sawfly population dynamics: patterns, processes, problems. Oikos 50:327–335

Hanski I, Parviainen P (1985) Cocoon predation by small mammals and pine sawfly population dynamics. Oikos 45:125–136

Haukioja E, Neuvonen S, Hanhimäki S, Niemelä P (1988) The autumnal moth in Fennoscandia. In: Berryman AA (ed) Dynamics of forest insect populations: patterns, causes, and management strategies. Plenum Press, New York, pp 163–178

Hunter AF (1991) Traits that distinguish outbreaking and nonoutbreaking Macrolepidoptera feeding on northern hardwood trees. Oikos 60:275–282

Jepsen JU, Hagen SB, Ims RA, Yoccoz NG (2008) Climate change and outbreaks of the geometrids *Operophtera brumata* and *Epirrita autumnata* in subarctic birch forest: evidence of a recent outbreak range expansion. J Anim Ecol 77:257–264

Jepsen JU, Hagen SB, Hogda KA, Ims RA, Karlsen SR, Tommervik H, Yoccoz NG (2009) Monitoring the spatio-temporal dynamics of geometrid moth outbreaks in birch forest using MODIS-NDVI data. Remote Sens Environ 113:1939–1947

Jepsen JU, Biuw M, Ims RA, Kapari L, Schott T, Vindstad OPL, Hagen SB (2013) Ecosystem impacts of a range expanding forest defoliator at the forest–tundra ecotone. Ecosystems 16:561–575

Jönsson AM, Bärring L (2011) Future climate impact on spruce bark beetle life cycle in relation to uncertainties in regional climate model data ensembles. Tellus 63A:158–173

Juutinen P (1967) Zur Bionomie und zum Vorkommen der Roten Kieferbuschhornblattwespe (*Neodiprion sertifer* Geoffr.) in Finland in den Jahren 1959–1965. Comm Inst For Fenn 63:1–129

Juutinen P (1982) Vorkommen und biologische Bekampfung der Roten Kiefernbuschhornblattwespe (*Neodiprion sertifer*) in Finland. Allg Forstz 37:230–232

Kallio P, Lehtonen J (1973) Birch forest damage caused by *Oporinia autumnata* (Bkh.) (Lep, Geometridae) in 1965–66 in Utsjoki, N Finland. Rep Kevo Subarctic Res Stn 10:55–69

Kingsolver JG, Woods HA, Buckley LB, Potter KA, MacLean HJ, Higgins JK (2011) Complex life cycles and the responses of insects to climate change. Integr Comp Biol 51:719–732

Klemola T, Andersson T, Ruohomäki K (2008) Fecundity of the autumnal moth depends on pooled geometrid abundance without a time lag: implications for cyclic population dynamics. J Anim Ecol 77:597–604

Kopisto L, Virtanen T, Pekkanen K, Mikkola K, Kauhanen H (2008) Tunturimittarituhotutkimus Käsivarren alueella 2004–2007. Metlan työraportteja/Working Papers of the Finnish Forest Research Institute 76:1–24

Kouki J, Löfman S, Martikainen P, Rouvinen S, Uotila A (2001) Forest fragmentation in Fennoscandia: linking habitat requirements of wood-associated threatened species to landscape and habitat changes. Sand J For Res Suppl 3:27–37

Långström B, Lindelöw Å, Schroeder M, Björklund N, Öhrn P (2009) The spruce bark beetle outbreak in Sweden following the January-storms in 2005 and 2007. In: Insects and Fungi in Storm Areas, Proceedings Workshop of IUFRO Working Party 7.03.10, pp 13–19

Larsson S, Tenow O (1984) Areal distribution of a *Neodiprion sertifer* (Hym, Diprionidae) outbreak on Scots pine as related to stand condition. Ecography 7:81–90

Lempa, K, Neuvonen, S, Tømmervik H (2005) Sustainable reindeer herding in mountain birch ecosystem. Chapter 19 In: Wielgolaski F-E (ed) Plant ecology, herbivory and human impact in Northern Mountain Birch Forests. Springer Verlag, Ecological Studies 180:269–273

May RM (1976) Simple mathematical models with very complicated dynamics. Nature 261:459–467

Mikkonen S, Laine M, Mäkelä H, Gregow H, Tuomenvirta H, Lahtinen M, Laaksonen A (2015) Trends in the average temperature in Finland, 1847–2013. Stoch Env Res Risk A 29:1521–1529

Neuvonen S, Viiri H (2015) Varautuminen lisääntyviin hyönteistuhoriskeihin muuttuvassa maailmassa. Kasvinsuojelulehti 4(2015):108–111

Neuvonen S, Virtanen T (2015) Abiotic factors, climatic variability and forest insect pests. In: Björkman C, Niemelä P (eds) Climate change and insect pests. CAB International, Wallingford, pp 154–172

Neuvonen S, Niemelä P, Virtanen T (1999) Climatic change and insect outbreaks in boreal forests: the role of winter temperatures. Ecol Bull 47:63–67

Neuvonen S, Bylund H, Tømmervik H (2005) Forest defoliation risks in birch forest by insects under different climate and land use scenarios in northern Europe. Chapter 9. In: Wielgolaski F-E (ed) Plant ecology, herbivory and human impact in northern mountain birch forests. Springer Verlag, Ecological Studies 180:126–138.

Neuvonen S, Tikkanen O-P, Viiri H (2016) Neljä vuotta kansallista kirjanpainajaseurantaa – feromoniseurannan tulokset 2015 ja muita havaintoja. Luonnonvara- ja biotalouden tutkimus 32(2016):28–32

Nevalainen S, Lindgren M, Pouttu A, Heinonen J, Hongisto M, Neuvonen S (2010) Extensive tree health monitoring networks are useful in revealing the impacts of widespread biotic damage in boreal forests. Environ Monit Assess 168:159–171

Nevalainen S, Sirkiä S, Peltoniemi S, Neuvonen S (2015) Vulnerability to pine sawfly damage decreases with site fertility but the opposite is true with Scleroderris canker damage; results from Finnish ICP Forests and NFI data. Ann For Sci 72:909–917

Niemelä P, Rousi M, Saarenmaa H (1987) Topographical delimitation of *Neodiprion sertifer* (Hym., Diprionidae) outbreaks on Scots pine in relation to needle quality. J Appl Ent 103:84–91

Nilssen A, Tenow O (1990) Diapause, embryo growth and supercooling capacity of *Epirrita autumnata* eggs from northern Fennoscandia. Entomol Exp Appl 57:39–55

Öhrn P, Långström B, Lindelöw Å, Björklund N (2014) Seasonal flight patterns of *Ips typographus* in southern Sweden and thermal sums required for emergence. Agric For Entomol 16:147–157

Økland B, Bjørnstad ON (2006) A resource depletion model of forest insect outbreaks. Ecology 87:283–290

Økland B, Netherer S, Marini L (2015) The Eurasian spruce bark beetle: the role of climate. In: Björkman C, Niemelä P (eds) Climate change and insect pests. CAB International, Wallingford, pp 202–219

Olofsson E (1987) Mortality factors in a population of *Neodiprion sertifer* (Hymenoptera: Diprionidae). Oikos 48:297–303

Potter KA, Woods A, Pincebourde S (2013) Microclimatic challenges in global change biology. Glob Chang Biol 19:2932–2939

Pouttu A, Annila E (2010) Kirjanpainajalla kaksi sukupolvea kesällä 2010. Metsätieteen aikakauskirja 4(2010):521–523

Pouttu A, Silver T (2016) Pistiäistilanne: Yyterin tähtikudospistiäistilanne syksyllä 2015. In: Metsätuhot vuonna 2015. Nevalainen S & Pouttu A (eds). Luonnonvara- ja biotalouden tutkimus 32/2016: 8–27

Ruohomäki K, Virtanen T, Kaitaniemi P, Tammaru T (1997) Old mountain birches at high altitudes are prone to outbreaks of *Epirrita autumnata* (Lepidoptera: Geometridae). Environ Entomol 26:1096–1104

Ruosteenoja K, Jylhä K, Kämäräinen M (2016) Climate projections for Finland under the RCP forcing scenarios. Geophysica 51:17–50

Santonen T (2011) Mittarituhot pohjoisen Utsjoen alueella ja sen vaikutukset alueen kasvillisuuteen ja poronhoitoon [Winter moth outbreaks in northernmost Fennoscandia and effects to vegetation and reindeer husbandry]. Thesis, 35 pp. Jyväskylä University of Applied Sciences, Finland

Schelhaas M-J, Nabuurs G-J, Schuck A (2003) Natural disturbances in the European forests in the 19th and 20th centuries. Glob Chang Biol 9:1620–1633

Seppälä M, Rastas J (1980) Vegetation map of northernmost Finland with special reference to subarctic forest limits and natural hazards. Fennia 158:41–61

Soubeyrand S, Neuvonen S, Penttinen A (2010) Mechanical-statistical modeling in ecology: from outbreak detections to pest dynamics. Bull Math Biol 71:318–338

Stadelmann G, Bugmann H, Meier F, Wermelinger B, Bigler C (2013) Effects of salvage logging and sanitation felling on bark beetle (*Ips typographus* L.) infestations. For Ecol Manag 305:273–281

Tenow O (1972) The outbreaks of *Oporinia autumnata* Bkh, *Operophthera* spp (Lep, Geometridae) in the Scandinavian mountain chain and northern Finland 1862–1968. Zool Bidr Uppsala Suppl 2:1–107

Tenow O, Bylund H, Nilssen AC, Karlsson PS (2005) Long-term influence of herbivores on northern birch forests. Chapter 12. In: Wielgolaski F-E (ed) Plant Ecology, Herbivory and Human Impact in Northern Mountain Birch Forests. Springer Verlag, Ecol Studies 180:165–181

Varama M, Niemelä P (2001) Männiköiden neulastuholaiset [The defoliators in Scots pine forests] Metsät aikak 2/2001:275–279

Viiri H, Neuvonen S (2016) Kirjanpainajasta on tullut pysyvä ongelma Suomen kuusimetsille – Mitä olisi tehtävä? Kasvinsuojelulehti 2(2016):57–61

Viiri H, Ahola A, Ihalainen A, Korhonen KT, Muinonen E, Parikka H, Pitkänen J (2011) Kesän 2010 myrskytuhot ja niistä seuraava hyönteistuhoriski. Metsät aikak 3(2011):221–225

Virtanen T, Neuvonen S (1999a) Climate change and Macrolepidopteran biodiversity in Finland. Chemosphere Global Change Sci 1:439–448

Virtanen T, Neuvonen S (1999b) Performance of moth larvae on birch in relation to altitude, climate, host quality and parasitoids. Oecologia 120:92–101

Virtanen T, Neuvonen S, Nikula A, Varama M, Niemelä P (1996) Climate change and the risks of Neodiprion sertifer outbreaks on Scots pine. Silva Fennica 30:169–177

Virtanen T, Neuvonen S, Nikula A (1998) Modelling topoclimatic patterns of egg mortality of *Epirrita autumnata* (Lepidoptera: Geometridae) with a Geographical Information System: predictions for current climate and warmer climate scenarios. J Appl Ecol 35:311–322

Voolma K, Pilt E, Õunap H (2009) The first reported outbreak of the great web-spinning pine-sawfly, *Acantholyda posticalis* (Mats.) (Hymenoptera, Pamphiliidae), in Estonia. Forestry Studies/Metsanduslikud Uurimused 50:115–122

Wermelinger B, Seifert M (1999) Temperature-dependent reproduction of the spruce bark beetle *Ips typographus*, and analysis of the potential population growth. Ecol Entomol 24:103–110

Wielgolaski F-E, Karlsson PS, Neuvonen S, Thannheiser D, Tømmervik H, Gautestad AO (2005) The Nordic mountain birch ecosystem – challenges to sustainable management. Chapter 25 In: Wielgolaski F-E (ed) Plant ecology, herbivory and human impact in northern mountain birch forests. Springer Verlag, Ecological Studies 180:343–356

Williams CM, Henry HAL, Sinclair BJ (2014) Cold truths: how winter drives responses of terrestrial organisms to climate change. Biol Rev 90:214–235

Chapter 6
Wood-Based Energy as a Strategy for Climate Change Mitigation in the Arctic-Perspectives on Assessment of Climate Impacts and Resource Efficiency with Life Cycle Assessment

Laura Sokka

Abstract Northern countries are committing themselves to large cuts in the greenhouse gas (GHG) emissions within the next decades. For example, the EU has agreed to cut down its GHG emissions by 40% by 2030. In a similar manner, Norway has announced commitments to reduce its GHG emissions by 40% by 2030 compared to 1990. Achievement of these emission reduction targets will mean shifting the balance of energy consumption in the region towards renewable sources such as wind, solar and biomass. There are large forest resources in the Nordic countries. Moreover, as a result of warming climate, the boreal forest line is expected to move northwards, displacing 11–50% of the tundra by boreal forests within the next 100 years. Increasing the use of bioenergy can provide emission reductions while also simultaneously help to reduce regional reliance on fossil fuels. On the other hand, increased mobilisation of forest biomass for energy decreases the growth of forest carbon sink and may in some cases even turn it into a carbon source.

In the present chapter, the use of forest bioenergy to ensure energy security and climate change mitigation is discussed. In addition, conclusions are drawn on how to simultaneously enhance energy security and resource efficiency, and contribute to emission reduction.

L. Sokka (✉)
Fulbright Arctic Initiative, Helsinki, Finland
e-mail: Laura.i.sokka@gmail.com

© The Author(s) 2017
K. Latola, H. Savela (eds.), *The Interconnected Arctic — UArctic Congress 2016*,
Springer Polar Sciences, DOI 10.1007/978-3-319-57532-2_6

6.1   Introduction and Background

In order to efficiently limit climate change, drastic changes in our present energy systems are needed. Increased use of forest and other biomass for energy has been identified as one central measure for climate change mitigation (Matthews et al. 2014). Forests also have an important role as carbon sinks in the mitigation of climate change (Pan et al. 2011). Increased mobilization of forest biomass for energy decreases the growth of this sink and may even turn it into a carbon source.

While forests and other bioenergy are considered an important source of energy, there are also many other existing and new uses for biomass. In the Arctic, in addition to being a source of materials and fuels, forests have important roles for example for reindeer herding and recreational use, including berry and mushroom picking. Forests are also a central source of biodiversity, particularly the old-growth forests (Koponen et al. 2015). Therefore decisions on alternative forest use options have to be made under complex and uncertain conditions. This calls for comprehensive assessments.

In this study, the use of forest biomass for energy under these complex conditions is discussed. The problematics related to the climate impacts of the use of slow-rotating forests are presented. In addition, the multiple use of forest biomass and the possibilities to combine the different uses are discussed.

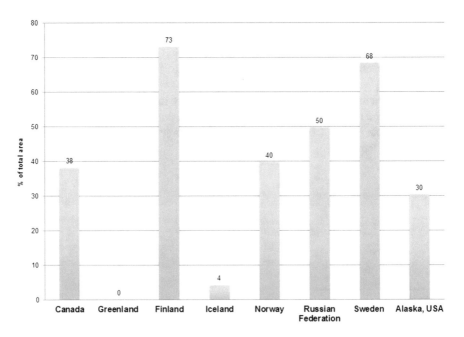

Fig. 6.1 Forest area as percentage of total area in the Arctic countries (FAO 2016)

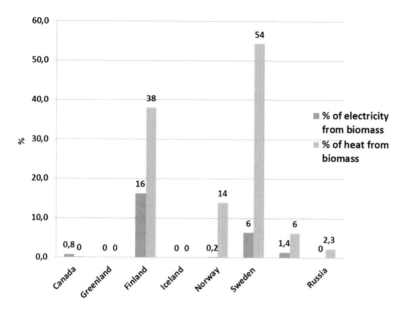

Fig. 6.2 Share of biomass of the total electricity and heat production (IEA 2015a, b)

6.2   Forests in the Arctic Countries

Many of the Arctic countries have large forest resources and forests play a substantial economic role in several of them (Fig. 6.1). For example, the USA, Canada, Sweden, Russia and Finland are among the world's largest exporters of pulp, paper and saw mill products (FAO 2015). In Norway, forest industry's role in the national economy is much smaller. Furthermore, in Alaska forest industry is fairly small and wood exports make up only a few percent of the State's exports (Alaska Forest Association 2016).

In some of the Arctic countries, forest biomass also plays an important role in the energy mix (Fig. 6.2). This is particularly true for Finland, Sweden and Norway. On the other hand, in the USA, Canada and Russia, the use of forest biomass ranges from zero to a few percent.

6.3   Climate Impacts Related to the Use of Forest Biomass for Energy

During the recent years, several studies have assessed the climate impacts of the use of boreal forests (for review see e.g. Matthews et al. 2014). The idea of the carbon neutrality of the use of forest biomass is based on the notion that in sustainable forestry, the extracted wood will eventually grow back, and re-absorb the carbon that was released (Helin et al. 2013). However, if wood is used for energy or for

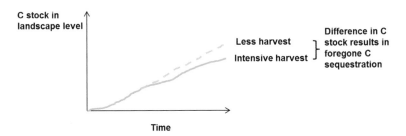

Fig. 6.3 Schematic figure of a difference between intensive and less intensive forest harvest scenarios (Figure adapted from Koponen et al. 2015)

other short-lived products, the carbon contained in it is released into the atmosphere quickly. As boreal forests are slow-growing, it takes from decades to even centuries for it to absorb the released CO_2 from the atmosphere. The so-called climate debt from the utilisation of forest biomass for energy, that has been discussed a lot during the past years, stems from the following: when biomass is taken from the forest, an unavoidable reduction in forest carbon stock is caused compared to a situation where biomass is not taken (Pingoud et al. 2015).

To conclude how efficiently biomass harvesting works in climate change mitigation, forest biomass harvesting needs to be studied in relation to a reference situation where less biomass is harvested for energy (see Fig. 6.3). In Fig. 6.3, two scenarios with different levels of extraction of forest biomass have been compared. The amount of carbon stored in the forest increases in both of them over time. Thus, forests form a carbon sink in both scenarios. However, in the case where forest is more intensively harvested, the resulting carbon stock is smaller. This impacts the atmospheric C balance. What the resulting actual total impact of the wood stemming from either one of these scenarios is, will depend on the use of the biomass, and what is replaced by it.

Life cycle assessment (LCA), which is a tool for quantitatively and systematically evaluating the potential environmental impacts of a product throughout its whole life cycle, has been applied in many studies that assess the climate impacts of wood biomass use (e.g. Holtsmark 2013; Mitchell et al. 2012). As in principle, in LCA all the inflows and outflows of substances, and the impacts of these, in a certain system, are assessed, LCA provides means to identify effective policy options. It also provides the kind of knowledge that reduces the risk of problem shifting. This implies e.g. situations where an improvement in one part of the life cycle leads to weakening in another time or place.

6.4 Adjusting the Different Uses of Forest Biomass

The on-going big initiatives for increased forest biomass use in the Finnish Lapland are driven first and foremost by climate change mitigation. However, climate change mitigation and energy industries are not the only users of forest biomass in the Finnish Lapland.

There are several different users of the forest biomass in the Arctic (e.g. Horstkotte et al. 2016). Even just in the Finnish Lapland, forests are used by forest and energy industry, reindeer herding, source of berries and mushrooms, tourism and even mining operations. Wood and forests are also influenced by policies focused on resource efficiency, energy, biodiversity, reindeer husbandry and tourism, among others.

Some previous research has studied the visions of the different stakeholders on forest biomass use [e.g. Horstkotte et al. 2016; Sténs et al. 2016; Lindahl 2015]. As could be expected, these studies indicate that the different groups have different visions of forest biomass use. However, synergies can also be identified. For example, increased energy use of wood implies increased self-sufficiency in energy use. Forests also have an important role as a source of employment in rural areas. This could mean, for example, rural-based small-scale entrepreneurship, ranging from the development of new wood products, berry and mushroom picking (as both subsistence use and for commercial purposes) and reindeer husbandry to tourism and recreation.

In Finland, there is a long tradition of policies concerning bioenergy production with specific recommendations for energy wood harvesting. Forests are also a key element in the national renewable energy policy, and increased use of bioenergy is also considered a potential way to improve the economic situation in the forest sector. Nevertheless, involvement of the local people in the decision-making through participatory methods and public hearings in e.g. environmental impact assessment is of central importance in order to implement climate change mitigation in the forest management planning (Ogden and Innes 2009). Furthermore, as different uses of forest biomass are often conflicting, management strategies that take into account the multiple values and uses have been found to be the best (Waeber et al. 2013).

Future research in the sustainable use of forest biomass for energy, particularly in the Arctic, should increasingly focus into identifying pathways that are sustainable from multiple perspectives. Understanding the impacts and challenges resulting from climate change need also further consideration.

References

Alaska Forest Association (2016) Alaska Forest Facts. Available in http://www.akforest.org/facts.htm. Cited on 10th Sept 2016

FAO (2015) Forest product consumption and production. Forest products statistics. http://www.fao.org/forestry/statistics/80938@180723/en/. Cited on 15th Jan 2017

FAO (2016) Faostat. Statistics Division. Food and Agricultural Organization. http://faostat3.fao.org/home/E Cited on 9th Sept 2016

Helin T, Sokka L, Soimakallio S, Pingoud K, Pajula T (2013) Approaches for inclusion of forest carbon cycle in life cycle assessment – a review. GCB Bioenergy 5(5):475–486. doi:10.1111/gcbb.12016

Holtsmark B (2013) Boreal forest management and its effect on atmospheric CO_2. Ecol Model 248:130–134

Horstkotte T, Lind T, Moen J (2016) Quantifying the implications of different land users' priorities in the management of boreal multiple-use forests. Environ Manag 57:770–783

IEA (2015a) Energy statistics of OECD countries. 2015 edition. IEA Statistics. International Energy Agency (IEA). Paris

IEA (2015b) Energy statistics of non-OECD countries, 2015 edition. IEA Statistics. International Energy Agency (IEA). Paris.

Koponen K, Sokka L, Salminen O, Sievänen R, Pingoud K, Ilvesniemi H, Routa J, Ikonen T, Koljonen T, Alakangas E, Asikainen A, Sipilä K (2015) Sustainability of forest energy in Northern Europe, VTT technology, 237. VTT, Espoo

Lindahl KB (2015) Chapter 8: Actors' perceptions and strategies: forests and pathways to sustainability. In: Westholm E, Lindahl KB, Kraxner F (eds) The future use of Nordic forests: a global perspective. Springer, Switzerland

Matthews R, Sokka L, Soimakallio S, Mortimer N, Rix J, Schelhaas M, ..., and Randle T (2014) Review of literature on biogenic carbon and life cycle assessment of forest bioenergy. Final Task 1 report (No. ENER/C1/427). The Research Agency of the Forestry Commission

Mitchell SR, Harmon ME, O'Connell KEB (2012) Carbon debt and carbon sequestration parity in forest bioenergy production. GCB Bioenergy 4:818–827

Ogden AE, Innes JL (2009) Adapting to climate change in the southwest Yukon: locally identified research and monitoring needs to support decision making in sustainable forest management. Arctic 62(2):159–174

Pan Y, Birdsey RA, Fang J, Houghton R, Kauppi PE, Kurz WA et al (2011) A large and persistent carbon sink in the world's forests. *Science 333*(6045):988–993

Pingoud K, Ekholm T, Soimakallio S, Helin T (2015) Carbon balance indicator for forest bioenergy scenarios. GCB Bioenergy 8:171–182. doi:10.1111/gcbb.12253

Sténs A, Bjärstig T, Nordström E-M, Sandström C, Fries C, Johansson J (2016) In the eye of the stakeholder: the challenges of governing social forest values

Waeber PO, Nitschke CR, Le Ferrec A, Harshaw HW, Innes JL (2013) Evaluating alternative forest management strategies for the Champagne and Aishihik Traditional Territory, Southwest Yukon. J Environ Manag 120:148–156

Chapter 7
Geospatial Analysis of Persistent Organic Pollutant Deposits in the Arctic Ecosystems and Environment

Vladimir A. Kudrjashov

Abstract The study was conducted to determine the distribution of accumulation of persistent organic pollutants (POPs) in the ecosystem components and environment in the Arctic regions of Russia. A GIS software was used to obtain data from regions and locations for which POP concentrations exceeded threshold values or remained within the normal range. The geospatial analysis was carried out for sea water, sea-bed sediments, and marine and terrestrial ecosystem components. The accumulation coefficients of POPs in the ecosystem components and environment were calculated. The obtained data analysis gave an opportunity to demonstrate the accumulation of POPs in the ecosystems. By data modeling it was determined that there is an exponential character of the POP accumulation in the components of Arctic ecosystems and environment.

7.1 Introduction

In the last decades, there has been a global growth in industrial activity, mining and agriculture intensification. Various pollutants are produced as a result of these processes. Among these pollutants are the persistent organic pollutants (POPs), such as DDT (dichlorodiphenyltrichloroethane), PCB (polychlorinated biphenyl) and HCH (hexachlorocyclohexanes), that remain active over a long period of time and also accumulate in the Arctic food chain in ecosystems and environment (Arctic Pollution 1998; AMAP 2004). POPs can have negative influence on ecosystems and environment, for example by disturbing the physiological functions of biota, polluting food chains, and by environmental contamination (Arctic Pollution 1998; Crane 2000;

V.A. Kudrjashov (✉)
Russian State Hydrometeorological University, Saint-Petersburg, Russia
e-mail: vakudrjashov@rambler.ru

© The Author(s) 2017
K. Latola, H. Savela (eds.), *The Interconnected Arctic — UArctic Congress 2016*,
Springer Polar Sciences, DOI 10.1007/978-3-319-57532-2_7

AMAP 2004). POPs are a large group of toxic organic substances. Chlorine ions are included frequently in the chemical composition of POPs as a one of their components. These compounds are named chlorinated hydrocarbons. Some of the POPs are direct or collateral products of industrial production, whereas other POPs have been produced or are produced as pesticides.

Transboundary transportation of trace amounts of POPs causes them to migrate into the Arctic from other parts of the Earth. Biomagnification effect is created due to the POP accumulation in the food chains of Arctic ecosystems (Arctic Pollution 1998; Crane 2000). By entering into polar biota, POPs can dissolve and accumulate in their fat tissues and then through the ecosystem chain by transfer from the lower parts of the food chains to higher ones with increasing POPs concentrations. Artificial organic chemical substances are very persistent to decompose in the arctic natural environment that is characterized by limited sunlight and lower temperature range. A forecast made with a help of mathematical model for a well-known pesticide, DDT decay indicated a decrease in pesticide mass from the initial concentration in air (10%), seawater (1–2%) and soil (30%) in ten years' time (Arctic Pollution 1998). POPs accumulated in the arctic biota are ingested with the traditional food of the indigenous peoples and can cause various physiological disturbances and diseases (Arctic Pollution 1998; Crane 2000; AMAP 2004).

7.2   Material and Methods

Geospatial analysis and modelling of contaminant concentrations of large regions provides an opportunity to (i) estimate the amount and location of pollutants, (ii) establish their dynamics and to (iii) predict POPs behavior in ecosystems and environment. This study focused on two toxic anthropogenic chlorinated hydrocarbons with heavy molecular weights: dichlorodiphenyltrichloroethane (DDT) and polychlorinated biphenyl (PCB). The geospatial analysis was made in the ArcGIS software environment.

The aims of the study included (i) quantitative evaluation of the two POP spatial distribution, (ii) comparison of their concentrations in the examined areas with threshold concentrations, (iii) calculation of two POP accumulation coefficients in the natural environment including sea water and sediments as well as in the components of arctic marine and terrestrial ecosystems. The geospatial analysis was carried out by using theme series of raster picture maps from the Arctic environmental atlas (Crane 2000), which were added as temporal grid-layers to the respective ArcGIS project. The threshold values indicating DDT and PCB values above the normal concentration range for environmental components and biota were obtained from tables in the Supplement B: Normal and threshold values of pollutant concentrations (Crane 2000).

Polluted habitats and regions were digitized manually into ArcGIS. Digitized polygon areas were calculated. The areas having equal pollutant concentrations were merged and calculated.

The conceptual model was created for estimation of the POP spatial impact in the Arctic ecosystems and environment. The model included an interactive histogram calculated on the basis of the merged polygon areas with the various POP concentrations. In the histogram, the abscissa axis shows the POP concentrations and the ordinate axis shows the merged polygon areas. The interactive histogram provided an opportunity to visualize and analyse the geospatial distribution of areas with various POP concentrations. A vertical red line was overlaid on the histogram. The line represents the threshold value of the POPs for various biota and environment components (Crane 2000). By using the red line as an indication, one can obtain the regions and locations for which the POP concentrations exceeded the threshold values or remained within the normal values.

The realization of the conceptual model for surface sea-bed sediment DDT pollution is presented in Fig. 7.1. Vertical red line on the interactive histogram shows the threshold value equal to 46 ng/g for the sediment (Crane 2000). All DDT concentration classes visualized on the interactive histogram did not exceed the threshold value.

Fourteen thematic vector data map layers of ArcGIS project were analysed by the conceptual model approach. The thematic data map layers contained areas polluted with DDT and PCB in various locations and concentrations. In the ArcGIS project, the following thematic data map layers were included for DDT: sea water, surface sea-bed sediments, seals (fat), beluga whales (fat). In the ArcGIS project, the following thematic data map layers were included for PCB: surface sea-bed sediments, seals (fat), beluga whales (fat), seagulls (liver), polar bears (fat), caribous

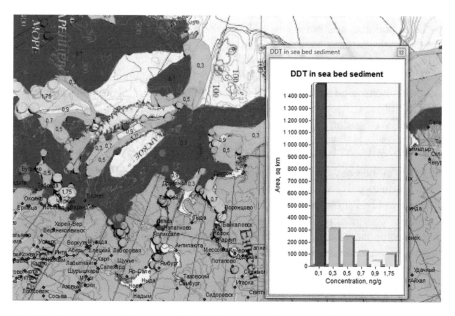

Fig. 7.1 The geospatial distribution of the DDT concentration in sea-bed sediments, based on the conceptual model

(liver), peoples (blood), peoples (breast milk fat). The analysis of DDT focused on two time periods of 10–15 years (1970–1985 and 1986–1996), providing an opportunity to compare the concentrations between the two time periods and to determine the pollutant concentration dynamics. The data for DDT and PCB concentrations were obtained from a series of raster picture thematic maps of the Arctic environmental atlas (Crane 2000) during its digitalization.

7.3   Results and Discussion

The results of the geospatial analysis revealed that a considerable part of the investigated areas had POP pollution. Nevertheless, only a small part of these areas had DDT or PCB concentrations that exceed the threshold values. Exceeding concentrations were identified in six of the investigated themes. These themes for DDT were sea water (two investigation periods) and beluga whales (fat). For PCB they were: seals (fat), beluga whales (fat) and polar bears (fat). Pollutant locations and areas of the given themes were determined and described. Comparison of the data obtained from the two investigation periods did not allow determination whether the POP accumulation was increasing or decreasing.

Subsequent advanced geospatial analysis included the calculation of accumulation coefficients for every POP in the biota components and environment. The accumulation coefficients were calculated with the formula: $K_a = C_f / C_p$, where K_a – POP accumulation coefficient, C_f – concentration of POPs in the following ecosystem level, C_p – concentration of POPs at the previous ecosystem level. The formula includes pollutant concentrations which are attributive values of geospatial polygonal GIS objects. They were calculated as an average POP concentration for every theme of investigation.

These pollutant concentrations, obtained from raster picture thematic maps of the Arctic environmental atlas (Crane 2000) were included into attribute tables of polygonal GIG objects. They were calculated as an average POP concentration for every theme of investigation or ecosystem components. The average POP concentration was calculated with the formula: $C_{ec} = \sum_{i=1}^{n} C_i / n$, where C_{ec} – average concentration of POPs in the ecosystem component, C_i – concentration of POPs in the merged thematic map polygon, n – the number of merged thematic map polygons. For example in Fig. 7.1 one can see the set of DDT concentrations in the surface sea-bed sediments presented on abscissa axes. The average concentration DDT in the ecosystem component i.e. in the surface sea-bed sediment is $C_{ec} = 0.71$ ng/g. The POP concentration values for geospatial analysis presented in Figs. 7.2, 7.3, 7.4, 7.5, and 7.6 were calculated by means of the described method and presented formula. The values of C_{ec} were shown near the data symbols or above diagram bars in the Figs. 7.4, 7.5, and 7.6.

Results of the K_a calculations are presented in Figs. 7.2 and 7.3.

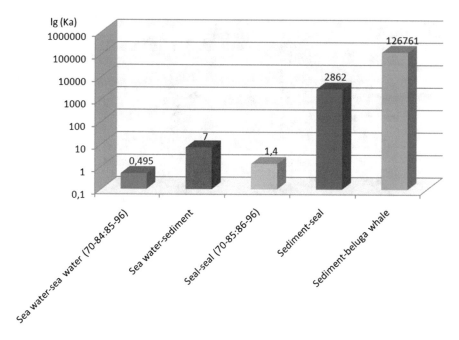

Fig. 7.2 Accumulation coefficients (K_a) for DDT in the marine ecosystem components. *Numbers above* bars represent K_a values

Accumulation coefficient values for DDT between the abiotic and biotic ecosystem components and between the two studied time intervals are presented in Fig. 7.2. Data analysis shows a well-detected trend of accumulation coefficient increase for transition from one abiotic component – sea water- to another component – sea bed sediment- and also from abiotic ecosystem components to biotic ones. The two time intervals on the DDT accumulation coefficient was not obtained in the sea water case, but K_a was obtained for the seal one.

The PCB accumulation coefficient values for marine and terrestrial ecosystem components are represented in Fig. 7.3. Absence of data for the PCB accumulation in the sea water is caused by the following factors: (i) PCB is accumulated in the water top layer on sea water – atmospheric boundary. The layer thickness varies from some microns to 1 mm. PCB concentration many times increases in the layer. (ii) In the sea water mass, PCB is absorbed by suspension and settles on the sea bed (Crane 2000). High accumulation coefficient values in the marine ecosystem components represented in Fig. 7.3 are observed in the transition from abiotic component to biotic ones and also in the biotic component itself. In the terrestrial ecosystem components, a considerable accumulation coefficient value was obtained when comparing the PCB concentration in the caribou liver tissue and in the breast milk fat.

More generalized analysis was carried out to obtain the quantitative characteristics and calculate equations connecting POP accumulations and Arctic ecosystem

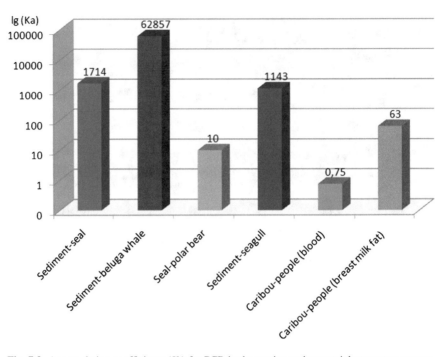

Fig. 7.3 Accumulation coefficients (K_a) for PCB in the marine and terrestrial ecosystem components. *Numbers* above *bars* represent K_a values

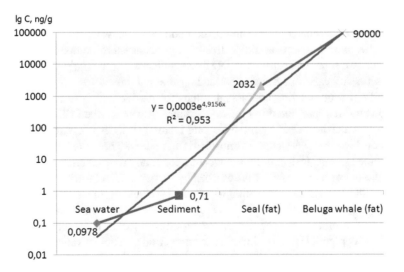

Fig. 7.4 The DDT concentrations in the marine ecosystem components. *Numbers* near the *symbols* represent the DDT concentrations (ng/g)

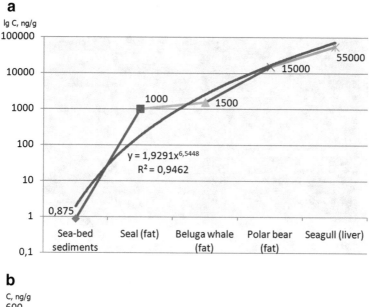

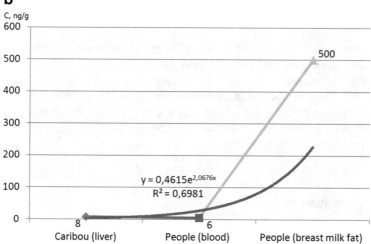

Fig. 7.5 PCB concentrations (**a**) in the marine ecosystem components, (**b**) in the terrestrial ecosystem components. *Numbers* near the *symbols* present the PCB concentrations (ng/g)

components. The relationships between POP concentrations and various ecosystem components and related equations are presented in Figs. 7.4 and 7.5.

The data analysis presented in Fig. 7.4 shows that DDT concentrations are sharply increasing in the ecosystem components. The increase in concentration is characterized by exponential dependence and is expressed in the equation presented in the figure.

Analysis of the PCB accumulation in the marine and terrestrial ecosystems components is presented in Fig. 7.5. PCB concentration shows an exponential increase

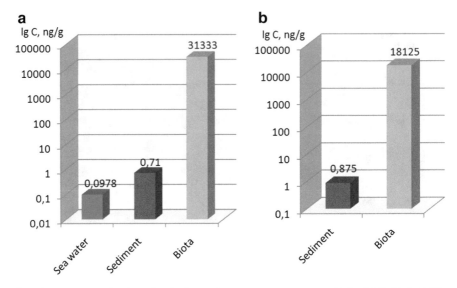

Fig. 7.6 POP magnification effect in the Arctic ecosystem components (**a**) for DDT, (**b**) for PCB. *Numbers* above the *bars* represent the POP concentrations (ng/g)

in the ecosystem components for both of the studied ecosystem. Related exponential equations are presented for both ecosystem components.

The magnification for DDT and PCB in the biota and environment is presented in Fig. 7.6. The average pollutant concentration in the sea water, sea-bed sediment and biota were used for the diagrams. The analysis of the diagram indicated that a considerable increase in the POP accumulation upwards in the hierarchical level of investigated ecosystems took place.

7.4 Conclusions

A geospatial analysis of the concentrations and distribution of two POPs, DDT and PCB, in the Russian Arctic regions showed, that these pollutants are present in most parts of the Russian Arctic, but that their observed concentrations do not exceed the threshold values in most parts of the investigated territories. Furthermore, it was determined that POP accumulation showed an exponential increase in the studied Arctic ecosystem components and environment, and that the magnification effect of the POP accumulation took place in the Arctic ecosystem components and environment.

References

AMAP (2004) Assessment 2002: persistent organic pollutants in the Arctic. Oslo, AMAP
Arctic Pollution (1998) Report about Arctic Environment State. AMAP. Hydrometeoisdat, Saint-Petersburg
Crane K (2000) Arctic environmental atlas. VNII Okeangeologia, WWF-USA and WWF Russian Programme Office

Chapter 8
Hydrological Probabilistic Model MARCS and Its Application to Simulate the Probability Density Functions of Multi-year Maximal Runoff: The Russian Arctic as a Case of Study

Elena Shevnina and Ekaterina Gaidukova

Abstract Climate warming has been and will continue to be faster in the Arctic compared to the other domains of the world, which generates major challenges for human adaptation. Among others, the development of socio-economic infrastructure and strategic planning requires long-term projections of water availability and extreme hydrological events. In this context, it is preferable that the projections of river runoff should be performed statistically, allowing the evaluation of economical risks and costs for hydraulic structures, which are connected to changes in hydrological extremes. In this study, the hydrological model MARCS (MARcov Chan System) is suggested as a tool to simulate the parameters of probability density functions (PDFs) of maximal runoff or peak flow, based on climate projections of the Representative Concentration Pathways. Following that, the PDFs of the maximal runoff were constructed within the Pearson Type III distributions to estimate the runoff values of a small exceedance probability. To evaluate the risks and costs of a long-term investment based on the future projections of river maximal discharge of 1 % probability, simple calculations were performed for the new bridge over the Nadym River as an example.

E. Shevnina (✉)
Finnish Meteorological Institute, P.O. Box 503, FI-00101 Helsinki, Finland

Russian State Hydrometeorological University,
Malookhtinsky pr., 98, RU-195196 Saint-Petersburg, Russia
e-mail: elena.shevnina@fmi.fi

E. Gaidukova
Russian State Hydrometeorological University,
Malookhtinsky pr., 98, RU-195196 Saint-Petersburg, Russia

© The Author(s) 2017
K. Latola, H. Savela (eds.), *The Interconnected Arctic — UArctic Congress 2016*,
Springer Polar Sciences, DOI 10.1007/978-3-319-57532-2_8

77

8.1 Introduction

Floods are among most dangerous natural hazards. Flood extremes cause substantial economic losses due to damage an infrastructure, such as bridges, roads, pipelines, dams, houses etc. Floods are commonly required in risk analysis of hydraulic structures (Scott 2011; Bowles 2001). Traditional methods of hydrological engineering use two main approaches to evaluate floods in given location at stream (Calver et al. 2009). The deterministic approach (or design storm method) combines a synthetic storm rainfall with physically-based hydrological model to evaluate the flood hydrograph (Pilgrim and Cordery 1975). The precipitation input is usually calculated from the intensity duration frequency curve of rainfall using observations in this case. However, the meteorological forcing can also be obtained from the climate projections, and then used to simulate the future river runoff hydrographs for a single or set of catchments (Archeimer and Lindström 2015; Lawrence and Haddeland 2011). The maximal discharges of particular exceedance probability are then evaluated from the modeled time series. The main short comings of this approach are that (i) the resulting flood estimations are sensitive to the algorithms of calculation of the projected meteorological values (Verzano 2009); (ii) the uncertainty produced by model parameters and algorithms can be significant (Butts et al. 2004); (iii) the calculations by the physically-based hydrological models are extremely costly computationally, especially in the case of ensembles of climate scenarios or regional scale assessment.

The statistical approach (or frequency analysis) considering multi-year runoff time series as realizations of stationary ergodic process, that can be performed statistically with distributions from the Pearson System (Andreev et al. 2005) fitted to the observations (Bulletin 17B 1982; SP33-101-2003 2004). The parameters of these distributions are evaluated from the observed runoff time series with different methods (van Gender and Vrijing 1997), and then the runoff quantiles of a small exceedance probability are estimated to support a design of the hydraulic structures. The greatest limitation of this method is that the runoff records should be representative for evaluation of the tailed values of fitted distributions (Katz et al. 2002). Also, the standard flood frequency analysis is based on the hypothesis of stationarity: it is assumed that the runoff statistics do not change over time, and past observations are representative for the future. However, the assumption of stationarity might not be applicable in changing climate, and advanced methods to evaluate flood frequency are required (Madsen et al. 2013).

To extend the statistical approach in the evaluation of flood and drought extreme quantiles under the expected climate change, the concept of quasi-stationarity of future hydrological regime is proposed by Kovalenko (1993). This idea allows evaluating the parameters of distributions of multi-year runoff for a future time period based on the sampled runoff statistics from past time period, and the distributions' parameters are different for the past (reference) and the future (projected) period. The climate projections are usually presented as the multi-year means of the meteorological values for the period of 20–30 years (Pachauri and Reisinger 2007), i.e. under the same assumption of quasi-stationarity. To simulate the future parameters of distributions of multi-year runoff based on quasi-stationary climate projections, the Kolmogorov Backward Equation (KBE) is simplified due to the specific of hydrological data and stationarity/quasi-stationarity concepts of engineering hydrology (Kovalenko 1993, Dominges and Rivera 2010, Viktorova and Gromova 2008).

In this study, the simplification of the KBE suggested by Kovalenko et al. (2010) was used, and a probabilistic hydrological model MARCS (MARcov Chan System) was developed as a tool using the Python capabilities. Recently, the MARCS model contains the following blocks for the preliminary analysis of runoff/meteorological time series, the model's parameterization and hindcasts, the forecasting based on the future climatology, the visualization and analysis of the modeling results.

The aims of this study were to describe the MARCS model structure and basic principles, and to provide an example of the model output in form of probabilistic projections of the maximal runoff for the Russian Arctic. Also, a method to utilize the projections of flood extremes to evaluate the economical investment to construct new bridges is suggested.

8.2　Method and Data

The probabilistic hydrological model, MARCS, includes five blocks. Each block contains the tools for (i) the data analysis and generalization (ii) the statistical methods for the data screening (both meteorological and hydrological), (iii) the schemes of the model parametrization (including regional oriented) and hindcasts, (vi) the forecasting, (v) the methods to perform statistical analyses and visualization of the modeling results (Table 8.1).

To simulate the probability density functions of multi-year maximal runoff over the territory of the Russian Arctic, the following data were used: the reference climatology was calculated from the meteorological time series, which were obtained from the catalogs of climatology for 209 weather stations (Radionov and Fetterer 2003, Catalogue 1989). The observed daily time series of river discharges for the period from early 1930s to 2002 from the R-ArcticNET database (www.r-arcticnet.sr.unh.edu/v4.0/) as well as the dataset of the Russian State Hydrological Institute (www.hydrology.ru/) were collected for 108 river gauges located over the northern Russia. The daily discharges were used to calculate the annual time series of the spring flood depth of the runoff (SFDR, mm/(time period)) as the volume of spring flood (m^3) from the drainage basin divided by its area (m^2). The volume of spring flood (m^3) was evaluated using flood beginning and ending dates (Shevnina 2013). The SFDR time series were used as the predicting value in the further analysis and modeling. The reason why the SFDR was chosen instead of the maximal discharge, is that it allows the regionalization and mapping, since the discharges depend on the watershed area. The river maximal discharge (m^3s^{-1}) with the required exceedance probability is calculated as described in Shevnina et al. (2016).

The core of the MARCS hydrological model is the Pearson System, which allows to model the PDFs within 12 types depending on the values of the parameters. The four parameters of the Pearson system could be estimated from the sampled initial

Table 8.1 The algorithms implemented to the MARCS model

MARCS unit	Algorithm	Purpose	References
Data Preparation Block (DPB)	Regularized spline with tension interpolation	To interpolate mean values of precipitation and air temperature for the reference/projected periods	Hofierka et al. (2002)
	Local minimums	To calculate yearly time series of SFDR using daily river discharge	Shevnina (2013)
	Delta correction method	To correct projected climatology from climate models' outputs	Hamlet et al. (2010)
Data Screening Block (DSB)	Spearman's Rank-Correlation Test	To analyze the runoff time series to the absence of trends	Zwillinger and Kokoska (2000)
	Fisher and Student Tests	To analyze the runoff time series to the stability of variance and mean	Fagerland and Sandvik (2009)
Parametrization and Hindcasts Block (PHB)	Floating period	To define the periods with statistically significant difference in the mean and variance, trends	Shevnina et al. (2016)
Model Core Block (MCB)	Basic parameterization	To calculate the model parameters from the observed runoff statistics and reference climatology	Kovalenko (1993)
	Regional parameterization for the Arctic	To calculate the model parameters from the observed runoff statistics and reference/projected climatology	Shevnina (2011)
Visualization and Analysis Block (VAB)	Uncertainties thresholds	To outline the regions with substantial changes in the mean value and coefficient of variation of the spring flood depth of runoff	Shevnina et al. (2016)
Economic Application Block	Modified cost-lost model	To evaluate total investments to a new bridge	Present study

moments (Stuard and Ord 1994). On the other hand, the parameters of the Pearson system can be evaluated from the parameters of KBE (see details in Kovalenko 2014, Domínguez and Rivera 2010, Pugachev et al. 1974). In our study, the Pearson Type III distribution were used to model PDF of SFDR and to evaluate the maximal discharges with required exceedance probability, thus the parameter denoted as b_2 in Andreev et al. (2005) was equal to 0.

The future climatology can be evaluated from the historical and projected runs of any Atmosphere-Ocean General Circulation Model (AOCGM) from the collection of the Coupled Model Intercomparison Project Phase 5 (CMIP5, http://cmip-pcmdi.llnl. gov/cmip5/). In the present study, the HadGEM2-ES climate model (Jones et al. 2011) was considered as an example. The projected mean values of the precipitation and air temperature for the period of 2010–2039 were used to force the MARCS model, and to simulate the probabilistic projections of the peak flow expressed though the PDF parameters of the spring flood depth of runoff. The model also allows calculating the maximal discharge of required exceedance probability at chosen site. In the present study the calculations was done for the Nadym River as described in Shevnina (2014).

In providing the practical example how the probabilistic projections of the peak flow can be applied in the economic, the evaluation of the total investments required for a new bridge was provided. In design of the hydraulic constructions the bridge height, type and costs are usually estimated using the peak flow discharges with particular exceedance probability (Megahan 1977, Guideline 1974). In this study the method to calculate the total investment to a bridge $(P, ₽)$ from the Guideline (1974) was modified by including two components; the possible losses due to flooding during operational period $(\alpha, ₽)$ and the additional investment to a bridge allowing decreasing risks of damaging due to flooding during operational period $(\beta, ₽)$ were added:

$$P = K + \beta + \alpha + \sum_1^t \frac{Э_t}{(1-E)^t} \qquad 8.1$$

where K is the total building cost $(₽)$; $Э_t$ is the cost during the operational period $(₽)$; t is the duration of the operational period equal to 35 years; E is the coefficient of inflation ($₽$ per year). Both additional components (α and β) are connected to the risk of accident due to extreme flooding during the period of operation of a bridge (*i.e.* 2010–2039). In present study, extreme flooding was considered as increasing the river flow the maximal discharge of 1% exceedance probability evaluated from the observed time series (namely reference 100-year discharge, Q_{IR}). In this study, the river maximal discharge of 1% exceedance probability (namely projected 100-year discharge, Q_{IP}) was also evaluated from the MARCS model based on the RCP4.5 climate projection. Thus, there are two options to estimate the total investment to the bridge depending of the height of structure based on the reference or projected 100-year discharges.

Considering the case when the Q_{IP} is equal to Q_{IR} or less then Q_{IR}, the total investment (P_1 in Fig. 8.1) contains only two components (the building and operational costs, K_1 and $Э_t$ in Fig. 8.1). It is expected that the risks to get any losses connected to flooding during operational period are low ($\alpha=0$) and there is no need for additional investments connected to the flood protection ($\beta=0$). In case when the Q_{IP} is larger then Q_{IR}, the risks to get temporal breaks in traffic or damage accident during a bridge operational period are also larger. In this case, the investment includes the expenses connected to an accident (P_2 in Fig. 8.1 in assumption that K, E and $Э_t$ are the same as in the previous case), and these losses are not equal to zero (α in Fig. 8.1). In this case, it is suggested to use the Q_{IP} to design of a bridge structure, then the additional costs (β in Fig. 8.1) are required. However these expenses may be less than losses due to the extreme flooding event. In this study, the total investments to the new bridge over the Nadym River at the Nadym City were calculated using reference and projected 100-year discharges. The total investments to the bridge (P_1 in Fig. 8.1) is equal to 14 milliards of rubles (see, www.finmarket.ru/news/4107593); in this study the additional components take amount of 10% and 2 % of K_1 for loses α and cost β respec-

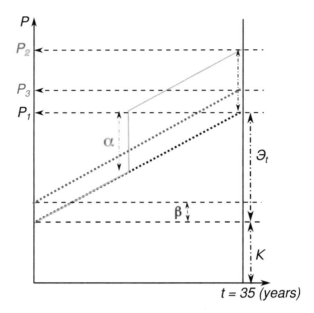

Fig. 8.1 The estimation of the total investment to a bridge construction by the Eq. (8.1)

tively. The projected values of 100-year discharge were calculated based on the outputs of four climate models under RCP 4.5 for the period of 2010–2039.

8.3   Results and Discussion

To evaluate the long-term projections of floods over the Russian Arctic, the MARCS model were developed as the tool, and it now contained four blocks. The Data Preparation Block (DPB) implemented the tools to evaluate the reference climatology from the observations at weather stations or gridded datasets, and the projected climatology from climate model outputs using the delta correction method (Hamlet et al. 2010). To map the reference/projected climatology, the regularized spline with tension interpolation technique by Hofierka et al. (2002) was implemented. The DBP also included the procedures to calculate the SFDR using the observed daily river discharges (Shevnina 2013).

The Data Screening Block (DSB) contained the tools implementing the statistical screening for the absence of trends and the stability of variance and mean (Dahmen and Hall 1990). In the screening procedures, the time series were divided into two sub-series by known year, when a distortion of a natural regime is expected (*i.e.* due to building a reservoir, starting a water regulation, water withdrawals, *etc*). However, the year of subdivision is usually not known, thus two algorithms were implemented into the model. In the "floating point" technique, the yearly time series were divided into two sub-series, and the lengths of the sub-series were usually different. Therefore, the method suggested by Fagerland and Sandvik (2009) was implemented to evaluate the values of Fisher and Student Test for sub-samples with unequal sizes. The "floating period" technique of the time series division uses two sub-series with equal size and is described in details in Shevnina et al. (2016). The significance of trend and the stability of variance and mean were checked on the statistical level equal to 0.05 in both algorithms. The procedures to calculate the sampled initial statistical moments based homogeneous runoff/meteorological time series were also implemented into the DSB.

The Parametrization and Hindcasts Block (PHB) included the procedure to evaluate the MARCS model hindcasts based on the runoff PDF parameters for two time periods using a cross-validation technique (see details in Shevnina et al. 2016). Two parameterization schemes were implemented into the MPHB: the general parameterization scheme and the regional parameterization scheme. In general scheme, the model parameters are constant for the reference and projected time periods (Kovalenko 1993). The regional parameterization scheme incorporates the projected climatology into the model parameters (Shevnina 2011).

The Model Core Block (MCB) contained the procedures to calculate the parameters of runoff PDFs based on the projected climatology (see details in Kovalenko et al. 2010). The forcing variables are the projected mean values of the precipitation and air temperature, and the output variables are mean value, the coefficients of variation and skewness of the runoff. The procedure to calculate the peak flow discharges with required exceedance probability using the Pearson Type III distribution was also implemented in this block.

The Visualization and Analysis Block (VAB) contained the tools to map the projected the mean value and coefficient of variation, and the regions with substantial changes in the runoff statistics (and extremes) compared with the reference period. Figure 8.2 shows the regions with changes in the SFDR means (top) and in the coefficient of variation (bottom) for the territory of the Russian Arctic under RCP4.5 climate scenario. These regions were obtained by applying the thresholds of the simulation uncertainties (see Shevnina et al. 2016 for details). The yellow and red areas in Fig. 8.2 (top) indicate the regions, where the risk to get flooding increases in the future compared to the reference period. Thus, correcting for the maximal river discharges of small exceedance probability for the expected climate changes was suggested here to evaluate the total investment to the hydraulic structures. The Nadym River is located in the region with increase of the mean values of SPDR up to 15–30 %, according to RCP4.5 climate scenario (Fig. 8.2), and the projected value of 100 year maximal discharge is larger than the reference.

The Economic Application Block (EAB) should be developed in the future and included the procedures to evaluate the economical values based on the runoff probabilistic projections. In this study, the algorithm to calculate the total investment to a new bridge was implemented (see section Data and Method) and was tested an example of the bridge over the Nadym River at the Nadym City (Shevnina 2014). The results show that the total investment to the Nadym River bridge could be less about 13 % in case of using the projected 100-year maximal discharges of 1% exceedance probability under the given assumption about the costs and losses in the Eq. 8.1. Using the climate-based projection of maximal runoff of the 1 % exceedance probability during the designing stage may lead to decrease of the total invest-

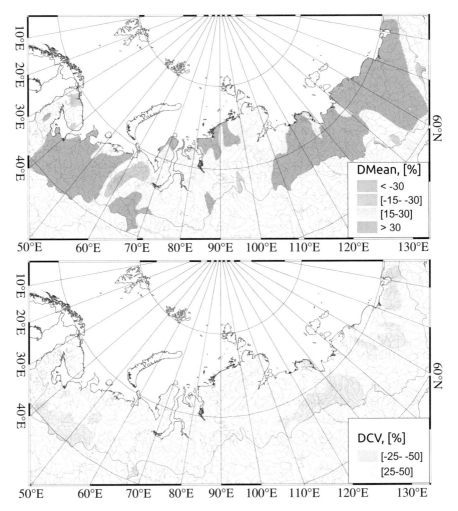

Fig. 8.2 The regions with substantial changes in the future mean values of the spring flood depth of runoff (*top*) and coefficients of variation (*bottom*) of the spring flood depth of runoff over the Russian Arctic (*red line* outlines its boundary) according to RCP 4.5 climate scenario: HadGEM2-ES model (2010–2039)

ment to the bridge construction over the Nadym River, which was already damaged by extremal flooding during the spring of 2016.

8.4 Conclusions

The main features of the MARCS model are (i) low numbers of the forcing and simulated variables (only the basic statistics of meteorological and hydrological variables are used); (ii) low numbers of the model parameters (the physical processes described integrally by the lumped model); (iii) the projected climatology is depicted by the model parameters, and it provides the way for developmet of the advanced regional-oriented parameterization schemes.

In this study, the MARCS model with the regional parameterization scheme was used to calculate the parameters of the SFDR PDFs over the Russian Arctic. However, the additional algorithms should be implemented to the model before developing regional-oriented schemes for the catchments located in the mid-latitudes. As the role of evaporation is more important in the general water balance on the south regions, this variable has to be considered (Kovalenko and Gaidukova 2011).

In this study, the MARCS model was forced by the outputs from the climate model with big spatial resolution, which is usually cause the challenges in the physically-based spatially distributed hydrological models. However, it seems to not be a big issue for the hydrological models such as the MARCS model. The physical core of the model is a lumped model in form of a linear filter with stochastic components (see Kovalenko 1993 for details). The role of spatial resolution in the datasets used to force the MARCS model and the options connected with the regional climate projections can be issues for future studies.

The vision for the future is changing continuously, and the set of climate change scenarios is renewed almost every 5 years since the meteorological models are improving unceasingly. The feature of the MARCS model is its general simplicity (only the projected statistics are evaluated instead of a time series), and is easy to perform for a regional scale assessment of the future water availability not only for an mean runoff value, but also for outliers (i.e for extreme hydrological events) under any chosen climate projection. These outliers are important for economists since they are usually the ones dealing with risks associated with weather/runoff extremes. The methods to evaluate the economic values from the outputs of the MARCS hydrological models are the topic of the future studies.

Acknowledgments The study was supported by the Academy of Finland (contract 283101, project TWASE) and by the Ministry of Science and Education of the Russian Federation (contract 01 2014 58678).

References

Andreev A, Kanto A, Malo P (2005) Simple approach for distribution selection in the Pearson System. Working paper of Helsinki School W-388. Helsinki

Archeimer B, Lindström G (2015) Climate impact on floods: changes in high flow in Sweden in the past and the future (1911–2100). Hydrol Earth Syst Sci 19(2):771–784

Bowles D (2001) Evaluation and use of risk estimates in Dam safety decision making. In: Risk-based decision making in water resources, IX. American Society of Civil Engineers, Reston, pp 17–32

Bulletin 17–B Guideline for determining flood flow frequency (1982) U.S. Geological Survey, Virginia, 72p

Butts MB, Payne JT, Kristensen M, Madsen H (2004) An evaluation of the impact of model structure on hydrological modeling uncertainty for streamflow simulation. J Hydrol 298(1–4):242–266

Calver A, Stewart E, Goodsel G (2009) Comparative analysis of statistical and catchment modeling approaches to river flood frequency estimation. J Flood Risk Manag 2:23–31

Catalogue of climatology of USSR Serie 3: multi-year data (1989) Gidrometeoizdat, Leningrad. (In Russian)

Dahmen ER, Hall MJ (1990) Screening of hydrological data: tests for stationarity and relative consistency. International Institute for Land Reclamation and Improvement, Dordrecht, 58p

Domínguez EK, Rivera H (2010) A Fokker–Planck–Kolmogorov equation approach for the monthly affluence forecast of Betania hydropower reservoir. J Hydroinf 12(4):486–501

Fagerland MW, Sandvik L (2009) Performance of five two-sample location tests for skewed distributions with unequal variances. Control Clin Trials 30:490–496

van Gender PHAJM, Vrijing JK (1997) A comparative study of different parameter estimation methods for statistical distribution functions in civil engineering applications. Struct Safety Reliability 1:665–668

Guideline on the economic calculations in construction and reconstruction of bridges VSN 2–73 (1974) The standards in building constraction, [online]. Available at: http://standartov.ru/norma_info/48/48033.htm. Accessed 1 Feb 2017. (in Russian)

Hamlet AF, Salathé EP, Carrasco P (2010) Statistical downscaling techniques for global climate model simulations of temperature and precipitation with application to water resources planning studies. Final Report for the Columbia Basin Climate Change Scenarios Project [online], 27 p. Available at: www.hydro.washington.edu/2860/products/sites/r7climate/study_report/CBCCSP_chap4_gcm_final.pdf. Accessed 1 Feb 2017

Hofierka J, Parajka J, Mitasova H, Mitas L (2002) Multivariate interpolation of precipitation using regularized spline with tension. Trans GIS 6(2):135–150

Jones C, Hughes J, Bellouin N, Hardiman S, Jones G, Knight J, Liddicoat S, O'onnor F, Andres R, Bell C, Boo K, Bozzo A, Butchart N, Cadule P, Corbin K, Doutriaux-Boucher M, Friedlingstein P, Gornall J, Gray L, Halloran P, Hurtt G, Ingram W, Lamarque J, Law R, Meinshausen M, Osprey S, Palin E, Chini L, Raddatz T, Sanderson M, Sellar A, Schurer A, Valdes P, Wood N, Woodward S, Yoshioka M, Zerroukat M (2011) The HadGEM2-ES implementation of CMIP5 centennial simulations. Geosci Model Dev 4:543–570

Katz RW, Parlange MB, Naveau P (2002) Statistic of extremes in hydrology. Adv Water Resour 25(8-12):1287–1304

Kovalenko VV (1993) Modeling of hydrological processes. Gidrometizdat, Sankt-Peterburg, 178p. (in Russian)

Kovalenko VV (2014) Using a probability model for steady long-term estimation of modal values of long-term river runoff characteristics. Russ Meteorol Hydrol 39(1):57–62

Kovalenko VV, Gaidukova EV (2011) Influence of climatological norms of the surface air temperature on the fractal dimensionality of the series of long-term river discharge. Dokl Earth Sci 439(2):1183–1185

Kovalenko V, Victorova N, Gaydukova E, Gromova M, Khaustov V, Shevnina E (2010) Guidelines on robust runoff estimation for projected hydraulic structures under non-steady climate. RSHU Publishing Co, St. Petersburg, 51 p. (in Russian)

Lawrence D, Haddeland I (2011) Uncertainty in hydrological modeling of climate change impacts in four Norwegian catchments. Hydrol Res 42(6):457–471

Madsen H, Lawrence D, Lang M, Martinkova M, Kjeldsen TR (2013) A review of applied methods in Europe for flood-frequency analysis in a changing environment. NERC/Centre for Ecology & Hydrology on behalf of COST, [online]. Available at: http://nora.nerc.ac.uk/501751/. Accessed 2 Feb 2017

Megahan WF (1977) Reducing erosional impacts of roads. In: Guidelines for watershed management. Food and Agriculture Organization, United Nations, Rome, pp 237–261

Pachauri RK, Reisinger A (eds) (2007) Synthesis report, contribution of working groups I, II and III to the fourth assessment report of the Intergovernmental Panel on Climate Change. IPCC, Geneva   https://www.ipcc.ch/publications_and_data_publications_ipcc_fourth_assessment_report_synthesis_report.htm. Accessed 2 Feb 2017

Pilgrim DH, Cordery I (1975) Rainfal temporal patterns design floods. J Hydraul Div 101(1):81–95

Pugachev VS, Kazakov IE, Evlanov LG (1974) Basics of statistical theory of automatic system. Mashinostroenie, Moscow, 400 p

Radionov VF, Fetterer F (2003) Meteorological data from the Russian Arctic 1961–2000. National Snow and Ice Data Center, Boulder, [online]. Available at: http://nsidc.org/data/docs/noaa/g02141_esdimmet/. Accessed 2 Feb 2017

Scott G (2011) The practical application of risk assessment to dam safety. Geo-Risk 2011:129–168

Shevnina E (2011) Parametrization of the probabilistic model to evaluate multi-year statistics of spring flow depth of runoff over the territory of the Russian Arctic. Scient Notes Russian State Hydrometeorol Univ 21:38–46. (in Russian)

Shevnina E (2013) Algorithm to calculate spring flood characteristics from daily discharges. Probl Arctic Antarctic 1(95):44–50. (in Russian)

Shevnina E (2014) Changes of maximal runoff regime in the Arctic. Const Unique Build Struct 7(22):128–141. (in Russian)

Shevnina E, Kourzeneva E, Kovalenko V, Vihma T (2016) Assessment of extreme flood events in changing climate for a long-term planning of socio-economic infrastructure in the Russian Arctic. Hydrological Earth System Science,doi: 10.5194/hess-2015-504, discussion paper (in review)

SP 33-101-2003 Guideline to estimate basic hydrological characteristics (2004) Gosstroy, Moscow, 234 p. (in Russian)

Stuart A, Ord J (1994) Kendall's advance theory of statistics, Distribution theory, vol 1. Edward Arnold, London

Verzano K (2009) Climate change impacts on flood related hydrological processes: further development and application of a global scale hydrological model, Reports on Earth system science 71. Max Planck Institute for Meteorology, Hamburg, 198 p

Viktorova NV, Gromova MN (2008) Long-term forecasting of characteristics of minimal river runoff discharges in Russia in case of possible climate change. Russ Meteorol Hydrol 33(6):388–393

Willmott CJ, Matsuura K (2001) Terrestrial air temperature and precipitation: monthly and annual time series, [online]. Available at: http://climate.geog.udel.edu/~climate/html_pages/README.ghcn_ts2.html. Accessed 2 Feb 2017

Zwillinger D, Kokoska S (2000) CRC standard probability and statistics tables and formulas. Chapman & Hall, New York

Chapter 9
Assessment of Atmospheric Circulation in the Atlantic-Eurasian Region and Arctic Using Climate Indices. The Possible Applications of These Indices in Long-Term Weather Forecasts

Mikhail M. Latonin

Abstract Polar air outbreaks from the Arctic can be categorically considered as extreme weather events because monthly temperature anomalies both in the Arctic and middle latitudes may exceed 20 degrees. In this study, it was found out that both the North Atlantic Oscillation and the Arctic Oscillation indices are not sensitive to the two completely different types of polar air outbreaks in terms of distinguishing them. The physical origins of polar air outbreaks were highlighted, and their classification was carried out. Based on this classification, a conclusion about the existence of the North Siberian anomaly was made. According to its many features, this anomaly can be treated as one more action center of the atmosphere. This finding has allowed us to introduce a new climate index, which is called as the Atlantic Arctic Oscillation index. This index allows us to identify the two types of polar air outbreaks with a high level of recognition probability.

An interrelation between the new climate index and temperatures in the investigated regions was analyzed. Summer season in the middle latitudes is becoming colder, while winter season in the Arctic is becoming warmer, and the Atlantic Arctic Oscillation index shows it.

One of the most important reasons of Arctic sea ice melting is related to the domination for the past 20 years of the second type of polar air outbreaks that cause high positive air temperature anomalies in the eastern sector of the Arctic. In contrast, during 1960s the first type of arctic air outbreaks prevailed.

M.M. Latonin (✉)
Russian State Hydrometeorological University, Saint-Petersburg, Russia
e-mail: weathermikel@gmail.com

© The Author(s) 2017
K. Latola, H. Savela (eds.), *The Interconnected Arctic — UArctic Congress 2016*,
Springer Polar Sciences, DOI 10.1007/978-3-319-57532-2_9

9.1 Introduction

Among the known climate indices that characterize the weather in the Atlantic-Eurasian region, the North Atlantic and Arctic Oscillations indices can be named. There are many articles, where the interrelations between different climate indices and weather-forming meteorological fields are considered, for example, (Pokrovsky 2007; Smith et al. 2016; Luo et al. 2016; Hurrell 1995). When speaking about Eurasia, the influence of the North Atlantic Oscillation is the most remarkable in Western Europe; already in European part of Russia it is essentially weaker and spreads to Siberia only during certain years. It is also known that the Arctic Oscillation is very closely interrelated with the North Atlantic Oscillation. In fact, Arctic and North Atlantic Oscillations are different ways for describing the same phenomena although this issue is not completely agreed upon by researchers.

9.2 Two Types of Polar Air Outbreaks

Retrospective data on the strongest polar air outbreaks to the European part of Russia for the last 30 years during the cold season were used to classify arctic air outbreaks. The dataset was provided by Vitaliy Stalnov who is the acting member of Russian Geographical Society and association of researchers "Forecasts and Cycles" (Stalnov n.d.).

Monthly anomalies maps from NCEP/NCAR reanalysis dataset (http://www.esrl.noaa.gov/psd/cgi-bin/data/composites/printpage.pl) and monthly sea ice concentrations anomalies from National Snow and Ice Data Center (ftp://sidads.colorado.edu/DATASETS/NOAA/G02135/) were used as an instrument to represent different cases of polar air outbreaks in order to classify them.

From the analysis of the archives, two absolutely different types of polar air outbreaks were identified.

Characteristic features of polar air outbreaks are identified in the best manner by the sea level pressure (SLP) anomalies fields presented in the Fig. 9.1.

The main difference between these two types of polar air outbreaks is the temperature regime and sea ice conditions in the Arctic. During the first type of polar air outbreaks, in the eastern sector of the Arctic negative temperature anomalies are observed, which leads to the increased sea ice concentration. During the second type, this situation is reversed.

As seen, the Northern Siberia and North Atlantic are key regions for the formation of polar air outbreaks in the Atlantic-Eurasian region. The conducted classification of polar air outbreaks has shown that there is the North Siberian anomaly, and, according to many features, it can be treated as the one more action center of the atmosphere.

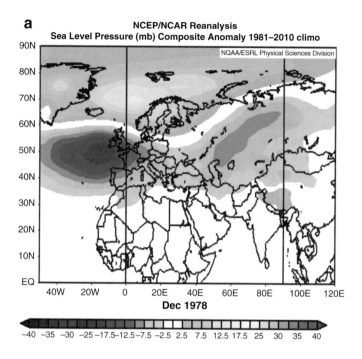

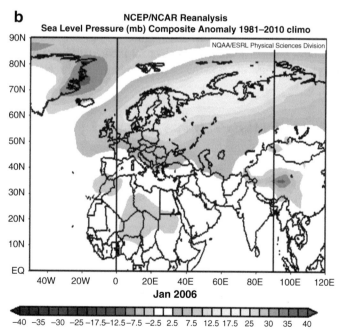

Fig. 9.1 (**a**) Example of the first type of polar air outbreaks in terms of SLP anomalies field, (**b**) Example of the second type of polar air outbreaks in terms of SLP anomalies field

9.3 Atlantic Arctic Oscillation Index (AAO Index)

9.3.1 Calculation of AAO Index

In the previous section, it was shown that there are two physical mechanisms that are inversely proportional to each other. At the same time, there are no climate indices for Central and Eastern Eurasia, which involve the action center of the atmosphere over the north of Siberia. In this view, a new climate index, entitled *Atlantic Arctic Oscillation*, was established.

For the calculation of this index, SLP data from two weather stations were used: Reykjavik (Iceland) and Ostrov Dikson (Russia). The international climate data were obtained from the Royal Netherlands Meteorological Institute (https://climexp.knmi.nl/selectstation.cgi?id=someone@somewhere). The missing values were filled up from NCEP/NCAR reanalysis dataset. In addition to the physical validity of the choice of these points, the Ostrov Dikson station has the longest time-series in the extreme conditions of the north. The derived index is the normalized difference of SLP anomalies between these stations. Normalization is used to avoid the series being dominated by the greater variability of the western station. To solve this task, a corresponding program was written by using the modern high-level programming language MATLAB.

Positive phase characterizes the second type of polar air outbreaks, whereas negative phase characterizes the first type.

In the Table 9.1, an example of the AAO index values is shown. Attention should be paid especially to January and June: distributions of positive and negative phases are nearly inversely proportional to each other for the two different periods of time. Further analysis will show in detail the physical meaning of these distributions.

9.3.2 Interannual Variability of AAO Index

The linear trend in Fig. 9.2 shows that there is a transition from negative phase to a positive one. However, the nonlinear smoothing, developed by O.M. Pokrovsky (Pokrovsky 2010), is much closer to the reality because it reveals a wave-like behavior of this oscillation, and due to this, e.g., it is possible to see that in January in 1960s the first type of polar air outbreaks dominated. Another very important tendency is that during the last 20 years the second type of polar air outbreaks has been prevailing.

According to the National Snow and Ice Data Center Report (Beitler 2012), the Arctic sea ice extent had its minimum value in 2012, which can also be explained by the domination of the second type of polar air outbreaks that has produced a regular transportation of very warm air in the Arctic. This, in turn, causes the sea ice not to thicken enough during winter, and the multiyear ice extent is gradually decreasing in summer. Therefore, it is possible to conclude that the impact of this

Table 9.1 Example of the AAO index values

Year	Jan	Feb	March	April	May	June	July	Aug	Sep	Oct	Nov	Dec
1963	-3.22	-2.16	0.40	0.06	3.01	0.32	-1.34	0.79	0.22	1.38	-0.96	-1.99
1964	-1.85	-0.16	-0.24	1.12	1.88	-0.91	2.12	-2.00	-0.56	0.20	-0.14	0.49
1965	0.01	-1.91	-1.31	1.89	-1.09	0.31	-2.16	-0.02	-0.66	-0.34	-0.78	0.84
1966	-1.01	0.14	0.87	2.54	-0.16	0.37	-0.76	-2.35	-2.05	-1.33	-0.56	3.08
1967	-2.04	-0.03	0.45	-1.20	-0.29	1.07	1.94	1.39	-0.82	-0.29	0.29	-1.78
2011	-0.09	1.17	-1.82	1.50	1.64	-0.80	-2.01	-1.48	3.73	0.77	0.65	0.83
2012	2.30	0.04	0.41	-0.66	-2.33	-0.39	-2.20	-1.57	1.52	0.53	1.14	1.21
2013	0.40	-0.95	-1.03	0.25	1.17	1.71	1.51	2.18	1.06	-1.67	-1.13	1.60
2014	2.40	2.85	-0.61	-1.46	-0.79	-1.15	1.13	-2.02	-0.96	2.46	-0.33	0.22
2015	0.61	-0.09	-0.13	-0.54	2.04	-1.16	-2.90	1.71	0.09	0.42	1.77	0.75
2016	1.76	0.25	-0.42	-0.90	0.04	0.57	0.77	-0.99	3.39	1.40	1.19	0.67

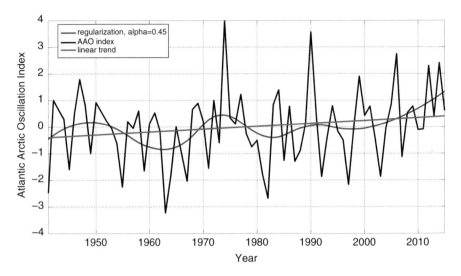

Fig. 9.2 Interannual variability of AAO index in January

process is cumulative, and as a consequence the year 2012 had the lowest ice extent since regular satellite observations started in 1979.

In June the situation is opposite, and currently the strength and number of polar air outbreaks of the first type is increasing. It should be pointed out that in summer the second type of polar air outbreaks vanishes in terms of lower than normal temperatures because of the high influence of solar radiation.

9.4 AAO Index and Surface Air Temperature in the Arctic. Advantages of the AAO Index Over the NAO and AO Indices

The most noticeable temperature anomalies in the Arctic are observed between the Svalbard and Frantz Josef Land. Therefore, it is interesting to assess the dependence of temperature on the AAO index concretely in this region. For this purpose, the time series of the mean surface air temperature (SAT) from NCEP/NCAR reanalysis dataset were taken.

It is clearly visible from Fig. 9.3 that in January the coherence between SAT and AAO is evident. The cross-correlation function confirms the visual conclusion. The correlation coefficient between the two time series is nearly 67%.

If every month of the year is considered, this results to the red curve presented in the Fig. 9.4. The curve shows that the maximum interconnection between the AAO index and SAT is observed in January, while in May the correlation is about zero. In

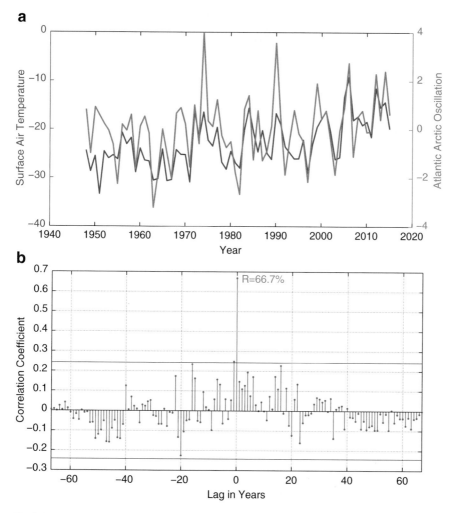

Fig. 9.3 (**a**) Interannual variability of AAO index and SAT in January, (**b**) Cross-correlation function for the sample of 1948–2015 years for January showing the interconnection of SAT and AAO index

general, the temperature regime during the winter season is well connected with the Atlantic Arctic Oscillation. During the summer season, the situation is vice versa.

In order to find out the advantages of AAO index compared to the two other indices, the same analysis was done with NAO and AO indices. The resulting green and blue curves, representing the interconnection of SAT with NAO and AO indices, are significantly below the red curve.

Another very important advantage of AAO index over NAO and AO indices is that it characterizes the two types of polar air outbreaks with a high accuracy.

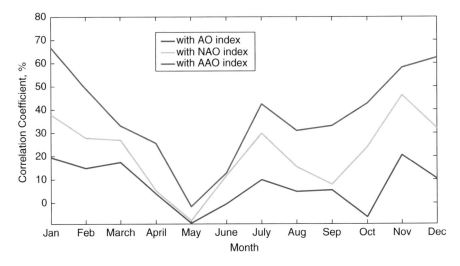

Fig. 9.4 Seasonal interconnection between SAT and three climate indices

Table 9.2 Sensitivity of three climate indices to the cases with polar air outbreaks obtained from the retrospective data (Stalnov n.d.)

	AO Index		NAO Index		AAO Index	
Month, year	I type	II type?	I type	II type?	I type	II type
December, 1978	−0.98		−2.34		−0.87	
October, 1979	−1.24		−0.67			**1.81**
March, 1980	−1.43		−0.67			**1.53**
January, 1982		**0.88**	−1.40		−2.68	
November, 1984	−0.97		−0.39			**0.98**
February, 1985	−1.44		−1.28			**1.31**
February, 1986	−2.90		−2.77		−2.62	
January, 2012	−0.22			**0.79**		**2.30**
January, 2016	−1.45		−0.37			**1.76**

Table 9.2 presents the numerical proof of AO and NAO indices nonsensitivity to the two completely different types of polar air outbreaks. The first column represents cases of polar air outbreaks (the bold font corresponds to the second type of polar air outbreaks; the ordinary font corresponds to their first type), and three others are monthly values of climate indices with the Atlantic Arctic Oscillation in the last column.

It can be seen from the table that Arctic and North Atlantic Oscillations are nearly of the same sign, whereas in reality the different types of polar air outbreaks were observed. Moreover, from the physical point of view, their positive phases even do not imply any arctic air outbreaks; therefore, question marks indicate that in the table for the second type of polar air outbreaks by NAO and AO indices.

In contrast to NAO and AO, Atlantic Arctic Oscillation is a very sensitive index to both types of polar air outbreaks. Its phases do not only tell us about the type of polar air outbreaks but also characterize the intensity of the corresponding anomalies. Physically this index reflects the strength of meridional atmospheric circulation, of which the arctic air outbreaks are a significant component. Therefore, there are more chances that this index will be successful in capturing the polar air outbreaks.

9.5 Conclusions

1. A new climate index, entitled *Atlantic Arctic Oscillation (AAO)*, was introduced in this study. AAO characterizes the two types of arctic air outbreaks with a high accuracy. The new index is much better interrelated with SAT in the Arctic than NAO and AO indices.
2. One of the most important reasons of Arctic sea ice melting is connected with the domination for the past 20 years of the arctic air outbreaks of the second type causing high positive air temperature anomalies in the eastern sector of the Arctic.
3. North Atlantic and Arctic Oscillations are not sensitive to the two completely different types of polar air outbreaks in terms of differentiating them.
4. Based on the conducted classification of polar air outbreaks, a conclusion was made about the existence of one more atmospheric action center over the north of Siberia.
5. Currently in the middle latitudes summer season is getting colder, whereas in the Arctic winter season is getting warmer. And the new climate index identifies this phenomenon.

References

Beitler J. (2012) Arctic sea ice extent settles at record seasonal minimum. In: Arctic Sea Ice News and Analysis. National Snow and Ice Data Center. http://nsidc.org/arcticseaicenews/2012/09/arctic-sea-ice-extent-settles-at-record-seasonal-minimum/

Hurrell J.W. (1995) Decadal trends in the North Atlantic Oscillation: regional temperatures and precipitation. Science 269:15–18

Luo D., Xiao Y., Yao Y., Dai A., Simmonds I., Franzke C.L.E. (2016) Impact of ural blocking on winter warm arctic–cold Eurasian anomalies. Part I: Blocking-induced amplification. J. Clim. 29

Monthly sea ice concentrations anomalies, National Snow and Ice Data Center, University of Colorado, Boulder. ftp://sidads.colorado.edu/DATASETS/NOAA/G02135/

Monthly sea level pressure data from weather stations, Royal Netherlands Meteorological Institute. https://climexp.knmi.nl/selectstation.cgi?id=someone@somewhere

Monthly/Seasonal Climate Composites, NCEP/NCAR reanalysis dataset. http://www.esrl.noaa.gov/psd/cgi-bin/data/composites/printpage.pl

Pokrovsky O.M. (2007) Primenenie dannykh distantsyonnogo zondirovaniya temperatury pover-khnosti okeana, ledovogo pokrova i atmosphery v Arktike dlya izucheniya tendentsyi izmen-eniya klimata Rossii (Implementation of Remote Sensing Data on the Sea Surface Temperature, Ice Coverage, and Atmospheric Parameters to Investigation of the Climate Change Trends in Russia). Issledovanie Zemli iz Kosmosa №2. pp 1–14

Pokrovsky O.M. (2010) Analiz faktorov izmeneniya klimata po dannym distantsyonnykh i kon-taktnykh izmerenii (The analysis of climate change factors according to the remote sensing and in-situ measurements). Issledovanie Zemli iz Kosmosa. №5. pp 1–14

Smith D.M., Scaife A.A., Eade R., Knight J.R. (2016) Seasonal to decadal prediction of the winter North Atlantic Oscillation: emerging capability and future prospects. Q. J. R. Meteorol. Soc. 142:611–617

Stalnov V. (n.d.) The strongest polar air outbreaks to the European part of Russia for the last 30 years during the cold season, Russian electronic journal "Meteoweb". http://meteoweb.ru/ar055.php

Chapter 10
Difficulties of Geological Engineering in Arctic Seas

Yuliia Tcibulnikova

Abstract The Arctic continental shelf is a promising area for oil and gas exploration and mining. Extremely harsh environmental conditions affect the work of engineering geologists, whose work is essential for building and construction of pipelines and rigs for the petroleum industry. With the massive interest and growth of fossil fuels offshore production, more and more geotechnical issues are to be solved. The necessity of studying marine sediments becomes clear when the specific physical and mechanical properties of bottom soils in the Arctic sea shelf are taken into consideration. Certain geological aspects determine what marine soils comprise and how they behave under loads exerted by a construction. Traditional methods of measuring deformation and strength parameters are reviewed, compared and contrasted by their feasibility of using to study marine sediments from offshore the Arctic. A substantial range of published studies has been analyzed and the findings summarized to provide potential solutions. The article stresses the importance of proper geotechnical survey and collaboration between industries and environmental scientists to achieve best results in studying the Arctic and building long-term human capacity alongside with protection of its vulnerable environment.

10.1 Introduction

Industries all over the world require an external energy supply. Up to the present days, the fossil carbohydrate fuels, such as oil and natural gas, have been the main energy resource, and despite of developments in alternative energy forms, the fossil fuels are still undoubtedly in huge demand. The petroleum exploration, development and mining has been ongoing for past century and a half. The sovereign states are pressing territorial claims in different areas of the Arctic Ocean, which Russian Federation tends to do as well, considering the Arctic continental shelf as a promising area for oil and gas industry.

Y. Tcibulnikova (✉)
St. Petersburg State University, Sankt-Peterburg, Russia
e-mail: juxik@yandex.com

© The Author(s) 2017
K. Latola, H. Savela (eds.), *The Interconnected Arctic — UArctic Congress 2016*,
Springer Polar Sciences, DOI 10.1007/978-3-319-57532-2_10

Harsh environmental conditions in the arctic sea areas make it rather difficult location to conduct economic activities that have recently been part of massive industrialization of the region in general. Both mounting vessels and petroleum industry constructions themselves must take the region's specific circumstances into consideration when planning long-lasting projects. The very fact that the sea bottom has to be reached introduces additional struggles to geological engineering. The object of studies (bottom sediments) lies several hundred meters deep under water, what makes it significantly hard to access. What is more, due to water entrainment the essential physical and mechanical properties of a soil may vary out of range of traditional testing methods. Even though ocean water masses downgrade the temperature variations, staying for long periods in the near-zero-degree water can be difficult to bear for divers who operate sea bottom sampling and drilling. Harsh environmental conditions also affect the vessel crews' performance of tasks. All the fore mentioned examples stress out the particular features and challenges of geological engineering offshore the Arctic.

10.2   Glimpse to the History of Petroleum Development Offshore the Arctic

Scientists first realized the necessity of studying sea bottom sediments in the 1930s. Until that time, the question had never been properly raised due to the absence of technologies allowing deep sea sampling. First ever continuous sequence of marine sediments was delivered by the German Meteor Expedition in the early 1930s. Scientists managed to determine the quaternary ice ages and global climate changes based on the data obtained. The expedition used so called the Kullenberg's piston corers that allowed taking unsplit samples of marine soils up to 20 meters depth (Kennett 1982).

When the GloMar Challenger research vessel was launched in 1968 in the United States, it had already been equipped with the echo-location positioning system. The high-accuracy scanning sonar captures the location of the initial borehole to continue drilling exactly in the same location in the next drilling session (Kennett 1982; DSDP 1969). After evidence of the Groningen gas province was discovered in Dutch part of the North Sea, in 1950s-1960s Norwegian petroleum scientists and entrepreneurs encouraged oil and gas exploration within the continental shelf of Norway (NP 2016). On the other side of the Arctic, Canadian government assumed the necessity to explore potential gas and oil deposits offshore the Northern Territories. In the late 1980s Canada granted licenses for petroleum activities in the Beaufort Sea and the McKenzie delta.

Offshore the Russian Arctic, only the Prirazlomnoye oil field is currently under the development. It was discovered in 1989 in the Pechora sea. Gazprom owns the production license for it. The extracted oil is in export demand on a massive scale. Theoretically it is possible to develop the Stockman gas condensate field in the

Barents Sea, which is currently suspended due to the shortage in financial support and the lack of required technological supplements.

10.3   Geological Engineering Conditions of the Region

Even though geological history, stratigraphic sequences and lithological features of marine sediments have been being studied for several decades (Rokos and Lyusternik 1992), the offshore soils required further in-depth research. Continental shelf is a submerged prolongation of the continent, featuring flattened underwater relief and geology complimentary to the adjacent firm land. Relatively shallow seas form above. For example, the average depth of Russian Arctic shelf seas is rarely more than 150 meters (Rokos and Lyusternik 1992; Kozlov 2012). The shelf areas only form by passive continental margins, where excessive tectonic deformations do not occur, where thick sedimentary cover can build up evenly during millions of years burying the oil and gas deposits. As studies from the Eastern Arctic area prove, the sedimentary cover has formed undisturbedly since the last stage of Caledonian orogeny circa 450 Ma, this being the average age of folded basement (Vinogradov et al. 2004). The sedimentary cover itself consists mainly of terrigenic sediments, altogether reaching the thickness of 17 km (Kozlov 2006). The desired fossil fuels are mainly found within the pore layers of those down to depths of 3 km.

The bottom offshore the Eastern Arctic is a spatial diversity of upheavals and thoroughs; the latter provide the deepest sea marking the greatest thickness of sedimentary layers. This in many cases is a key factor for forming fossil fuels deposits. One of the striking examples is the Eastern Barents sea depression, which contains the large oil and gas bearing Barentsevomorskaya province covering area of 60,000 km², and the Ludlovskoye gas condensate deposit (Kozlov 2010). However, gas has a tendency to escape from deeper major deposits (due to lower density or to the unconformities in covering sediments) and accumulate in upper layers of soil under clay lenses. When covering clay is disturbed during a rig construction, gas bursts out with considerable force comparable to an explosion, what can cause destruction of construction parts and mounting vessels (Kozlov 2010).

Most of the soils on the Arctic shelf are composed of mixed sand and clay particles. The upper layer of sea bottom sediments cannot be considered as a significant construction base due to its high transfer liability, low pressure resistance or unsolid bedding. The underlying layer of ground that is older in age (up to 2 Ma), is often used as an appropriate base for drilling rigs. However, another challenge is often caused by the presence of ice. The permafrost constitutes the major part of the area both on land and offshore as the result of paleoclimatic events such as the sequence of quaternary Ice ages (Rokos and Lyusternik 1992). This extremely stiff solid formation of water frozen in the pores of a soil gets exposed in large areas at the sea bottom. In addition, the upper layer of marine soil may contain local lenses of recently formed clear ice (Kozlov 2012). Both ice types pose significant challenges to construction. Permafrost is extremely hard to drill through, what makes building of the drilling rigs significantly difficult.

All the before mentioned conditions have direct implications on projecting oil rigs or pipelines within the Arctic shelf seas. To ensure the durability, safety of usage and stability of future structures as well as to prevent accidents, decent strength and deformation properties of the bottom soil have to fit the previously calculated ones of that structure. At this stage of geotechnical research, the environmental conditions of the region play a vital role. For example, even though standard methods can be used for moisture, consistency and porosity characterization, and for defining the grain size composition, the ocean water saltiness has to be taken into consideration. The salt component makes soil grains stick together into larger particles, highly prone to unpredictable crumbling under a structure pressure.

10.4   Issues on Testing

Since the upper layer of marine soils are very loose and highly watered, studying them involves the use of nonstandard methods. The basic soil studies include laboratory testing carried out on soil samples taken from the sea bottom. Contemporary practice uses various sampling devices, such as piston and vibration corers, box samplers, dredgers, snapper samplers. Operating all of the listed equipment is rather difficult and pricey, because their use requires qualified staff (Kennett 1982).

When the sediment sample is obtained from the sea bottom and brought undamaged to the laboratory, the actual geotechnical testing begins. For every type of soil it is needed to find the main mechanical parameters, which determine how the soil behaves under pressure exerted by a structure above. The parameters are divided into two major groups – the deformation properties (such as the porosity ratio) and the strength properties (the inner friction angle and the inner cohesion between grains). An oedometer shows the dependence of porosity ratio on exerted pressure. However, marine sediments are so loose that the critical pressure cannot be tracked. Either water is squeezed out, not meeting the natural condition of the soil (Rokos and Lyusternik 1992).

To test the strength properties of a soil, the friction angle and the inner cohesion of a sample are measured by exerting both vertical and horizontal sheering pressure to it. A box sheer apparatus and stabilometer are commonly used in testing an average firm land soil, whereas for marine soils those are hardly useful. Marine soil is so loose that the mentioned instruments are not precise enough to mark critical pressure (Ziangirov et al. 1982).

In marine soil studies, penetration probations are widely spread (Kennett 1982). The downside of this method is that it allows only the determination of the general sheer resistance. The precise values of friction angle and inner cohesion, directly used in projecting and construction, are hard to calculate with this methodology.

10.5 Potential Solutions for Arctic Geological Engineering

For marine sediments, especially those at the Arctic sea bottom, testing should be done *in situ* in the soil's natural conditions. Until recently, hiring qualified professional divers was essential to operate the measurement tools at the sea bottom. Nowadays, engineering geologists have opportunities to use multifunctional probes of various configurations to operate from the research vessel, which basically negates human error.

Concerning the challenges caused by permafrost, thermodrilling may prove its effectiveness, as long as the resulting excessive heat is reduced, and the surrounding ice is preserved from unwanted melting. This can be achieved by attaching a heat-and-water-resistant cover to the drill bit, preventing the heat from spreading into the ocean. As for the gas lenses inside a soil layer, when a lens is discovered, tapping (puncturing) it to let the gas pop out can be considered a solution. In this case, development and construction can continue as soon as the bottom soil appears relatively stable.

Taking into consideration all the listed particularities of marine soils offshore the Arctic, a detailed investigation and study of the area is required once it is licensed for petroleum development or construction. In the Arctic oceans and its coastal areas, the main branches of industry that are in the need of geological engineering data, are petroleum industry and related construction. Bottom soils offshore the Arctic have to be studied in cooperation with geophysicists to request complementary remote sensing data. The information provided by geological engineers is needed to build constructions for gas and petroleum industries. This requires collaboration with all the involved professionals in those fields. The environmental authorities' approval has to be received for any project in the Arctic region. The Arctic ocean and its coastal area are a unique ecosystems that are extremely fragile and vulnerable. An accident at pipelines or oil wells causes risks of irreparable damage not only to that specific ecosystem, but to the whole planet Earth.

References

Deep Sea Drilling Project (1969) DSDP initial report. International Ocean Discovery Program IODP

Kennett JP (1982) Marine geology. University of Rhode-Island/Prentice-Hall, Englewood Cliffs, USA

Kozlov SA (2006) Forming of geological Engineering conditions at the Barents-Kara shelf. Dissertation, VNIIOkeangeologia, St. Petersburg

Kozlov SA (2010) Conceptual bases of geological Engineering research of the Western Arctic petroleum province. VNIIOkeangeologia, St. Petersburg

Kozlov SA (2012) Geological issues in building gravity-based drilling platforms in Arctic seas. Min J 2012

Kozlov SA, Neizvestnov YV (2003) Spatial variability of the physical and mechanical properties of sea-floor sediments of the oil and gas-bearing area of the Barents-Kara shelf. In: Marine Engineering survey. VNIIOkeangeologia, St. Petersburg

Norwegian Petroleum (2016) Norway's Petroleum History. Norskpetroleum network electronic publication http://www.norskpetroleum.no/

Osadchiy A (2006) Oil and Gas offshore the Russian Arctic: assessment and projection. Science and Life 7/2006

Rokos SI (2008) Geological Engineering aspects of extrahigh reservoir pressure subsurface zones at the Pechora and South Kara shelf. Engineering Geology 4/2008

Rokos SI, Lyusternik VA (1992) Forming of composition and physical and mechanical properties of Pleistocene sediments in the southern and central Barents Sea shelf (genetic and paleogeographic aspects). Ukraine Academy of Science, Institute of Geological Sciences, Kyiv

Tarovik VI, Valdman NA, Verbitskiy SV, Zimin AD, Obidin YI (2012) Innovative solutions in offshore machinery. Krylov State Research Centre, Krylov Shipbuilding Research Institute, St. Petersburg

Vinogradov VA, Gusev EA, Lopatin BG (2004) Structure of the East Russian Arctic shelf. In: Geological-geophysical features of the lithosphere of the Arctic Region. VNIIOkeangeologia, St. Petersburg

Ziangirov RS, Root PE, Filimonov SD (1982) Soil mechanics practice. MSU, Moscow

Part II
Vulnerability of the Arctic Societies

Chapter 11
The Health Transition: A Challenge to Indigenous Peoples in the Arctic

Peter Sköld

Abstract Good health and well-being is one of the most important sustainability goals of today. Unfortunately the goal faces many challenges that show an uneven distribution of health improvements, and of life-expectancy. This is a global problem, but also a specific threat to vast parts of the Arctic. Furthermore there is a strong correlation between climate change and health risks. On top of these challenges are disfavoured indigenous peoples, globally and in the Arctic. This chapter deals with health encounters in the North, with a focus on the Swedish health care organization, Sami health and research efforts.

11.1 National Health Care Systems and Scientific Collaboration

Research plays an important role for the promotion of good health and well-being in the Arctic. Before any research comes naturally the development of a health care system. Historically this was of mere national concern for the respective Arctic countries, often the initiatives were not specifically targeted to the regions of the North. When Sweden established an organization with district physicians in the 1770s there was only a total of 32, and one single physician was responsible for the two northernmost counties, a huge area of responsibility that was impossible to cover properly. Over the nineteenth century the health care system slowly developed and reached out to remote areas too.

The emergence of the Swedish welfare state and its health system has been widely discussed in historical as well as sociological studies (Baldwin 1999; Porter 1999). On one hand Sweden was part of a European context, where scientific co-operation eliminated previously existing hinders. Swedish physicians had opportunities to study and work at foreign universities. This "correspondence" also meant that medical discoveries such as Jenner's smallpox vaccine quickly reached Sweden.

P. Sköld (✉)
Umeå University, Umeå, Sweden
e-mail: peter.skold@umu.se

© The Author(s) 2017
K. Latola, H. Savela (eds.), *The Interconnected Arctic — UArctic Congress 2016*,
Springer Polar Sciences, DOI 10.1007/978-3-319-57532-2_11

The public health strategies in most European countries developed a combination of public health focusing on both environmental and individual health policies. Preventive measures and public health propaganda (the hygienic movement) became key features, and gave Sweden a comparatively low health care expenditure, and there was a strong ambition to create a welfare state (Sundin and Willner 2007; Rehnberg 1990). As it happened the first national health and living conditions survey took place in northernmost Sweden 1929–1931. The so-called Investigation of the North involved 4400 physicians and dentists and 17,400 northerners, and concluded that the diet had to be improved with more vegetables, housing must improve, gardening be developed, and school medicine be established. It also noted that cardo-vascular diseases and alcoholism were relatively rare. One might argue that the efforts to create the Investigation of the North were important for the development of research at the hospitals generally, and to the establishment of Umeå University 35 years later.

The development of health care systems varied to a great extent over the Arctic. But all systems had to take on similar challenges dominated by large regions with long distances between people, a cold and often harsh climate, limited resources, and difficulties to recruit health professionals. International research collaboration was intensified and in 1967 the first Circumpolar Health Symposium was arranged in Fairbanks. When the scientific conferences were evaluated 35 years later, the conclusion was that three fields of research had dominated the conferences: epidemiology of indigenous peoples, health care delivery and effect of physical factors on human physiology and health (Bjerregaard et al. 2003). Arctic health conferences during the past decade reveal that the field has added strong initiatives on new technologies, digitization and e-health, living conditions, occupational health, climate, the spread of infectious diseases, indigenous health and mental health, and suicide problems to their activities.

Collaboration has also been a key strategy for the international Arctic research community. The three science organizations with observer status to the Arctic Council have all health on the agenda. The International Arctic Social Science Association (IASSA) has 20 health session at the ICASS IX conference in June 2017 and the International Arctic Science Committee has a working group (SHWG) that includes health issues and has supported initiatives on health statistics in the Arctic. University of the Arctic has a thematic network on health and well-being. And a great manifestation of Arctic health research is the International Congress on Circumpolar Health that has sixteen conferences behind them, last time arranged in Oulu, Finland 2015. IASSA, IASC and UArctic have signed a Letter of Agreement which increases the opportunities to work towards health sustainability, and research driven by solutions and social impact.

Sustainability, solutions, and social impact are important goals for health care in the Arctic. National and regional health care organization, and scientific collaboration have proved to be important for the implementation of the goals. Nevertheless, there are considerable challenges for good and equal health in the Arctic that remain. These are general challenges, even if there are great disparities in the region. This chapter deals with the health situation for the Indigenous peoples in the Arctic, highlighting the Sami.

11.2   The Sami People

The Sami is a native people living in an area covering parts of northern Norway, Sweden, Finland and Russia, in the Sami language known as Sápmi. Different opinions have been expressed in the efforts of creating the best circumstances for the Sami to develop their societies in the north. It is very important to stress that the Sami the challenge is not limited to preservation. The Sami societies are dynamic, and parts of perpetual change. The Sami yoik (traditional song) is today expressed in blues, jazz, hip hop and heavy metal contexts. Reindeer herding is very technical, which is necessary in times of economic rationalization and improved work conditions (Sköld 2015).

During the nineteenth century the idea that the Sami ought to be culturally incorporated to the Swedish nation and change their nomadic lifestyle was replaced by an ideology stating that the only way for them to survive was to remain isolated and unchanged. This political strategy was replaced by an assimilation policy, resulting in cultural stigmatization, defeat of Sami languages, and weakened ethnic identity. The last decades have, however, witnessed a cultural revitalization process paired with a better acceptance of a multi-cultural society in the north.

The concept of vulnerability is, to say the least, complex in a variety of meanings and contexts. Nevertheless, vulnerability is a relevant issue for the Sami. Not least indigenous people's vulnerability has been a hallmark throughout history. Hundreds, even thousands, of indigenous peoples have experienced vulnerability to such an extent that they today no longer exist. From the eighteenth century and onwards it has been said that the Sami people lives under the threat of extortion. Different opinions have been expressed in the efforts of creating the best circumstances for the Sami to exist as an Indigenous group in the area. A discussion of Sami vulnerability necessarily includes relatively large generalizations It is not just the vulnerability, but also the meaning of concepts such as culture, indigenous people and the Sami. Relevant for the Sami is the complexity that always has characterized their culture. Frequently used terms such as "the Sami want" and "the Sami believe" reveal the stereotype conceptions that have been present. We should remember that Sápmi crosses by a number of borders, cultural and national. There are nine Sami languages, which are divided into three main groups, and between these language groups there is a very limited linguistic understanding. There are South and North Sami. There are reindeer-herding Sami, forest Sami and Sea Sami. There are different groups in the four different countries including Sápmi. There are Sami in all counties and districts in Sweden. At the first elections to the Sami Parliament in Sweden, there were 17 different political parties represented. The Sami people do not like, and have never liked, the same thing. This cultural diversity within the Sami society has often been hidden to the general public.

11.3   Sami Health Challenges

Assuming that the complex Sami society is given account it can be argued that the Sami, in comparison with most indigenous peoples around the world, has a relatively good situation. The Sami in Sweden have experienced a unique and positive health development over the past centuries that has taken them from a very high mortality and low life expectancy to levels quite on par with the general average age in Sweden (Axelsson and Sköld 2006). In a recent article published in The Lancet (Anderson et al. 2016) poorer outcomes for Indigenous populations for life expectancy, infant mortality, maternal mortality, high birthweight, child malnutrition, child and adult obesity, educational attainment, and economic status are documented. The differences between regions and countries are great, but the Sami has a relative advantage to other Indigenous peoples.

Nevertheless, we should remember that reindeer herding is one of the most dangerous occupations with major accident hazards and that there are very worrying trends in high suicide rates among young reindeer herders. The Sami have increased their political influence. Since the early 1900s a relatively successful political mobilization, including the establishment of Sami Parliaments in Norway, Sweden and Finland, has occurred in the Sami society. Despite the advance in many areas the Sami are still dependent on decisions by the State and the majority society on most issues.

And there is still vulnerability in the Sami society. Reindeer herding, which is an important part of the Sami culture, are working under difficult economic and legal conditions. The Sami languages are fighting for its survival. We know that language and identity are closely linked. A stereotyped image of the Sami is still prevalent today and a general ignorance has been due to inadequate teaching and learning materials ignoring. This is an essential part of the Sami issues of today. Traditional knowledge and values threatens to be lost in a shrinking cultural space, while the Sami complexity and modernization has meant that more and more lose their Sami identity. And a crucial question is how much we really know about the Sami health situation.

11.4   Sami Health Data

It is uttermost important to have accurate information related to the health development in the Arctic. Generally, official registers present information about the inhabitants of the Arctic regions that is of equal quality compared to the non-Arctic parts of a respective country. There are, however, two major deficiencies; parameters that are compatible between the Arctic countries and data that has the capacity to illustrate the indigenous peoples separately.

Health is of course not equivalent to life-expectancy or mortality. Overall there is sufficient data for the Arctic regions to analyse the transition. But when it comes to methods and terminology aimed to cover life-quality, marginalization, discrimination, mental health, and living-conditions it is much more difficult to compare across nations. The Arctic countries also differ substantially in their efforts to include ethnicity. It is, however, often very difficult to trace ethnicity in both the historical records and in the present-day population statistics. Official and self-determined definitions have varied extensively over time, and between countries (ASI 2010).

The insufficient inclusion and categorization of ethnicity in registers creates difficulties to estimate not only population size and composition, but also specific features such as languages, education, occupation, and health status (Axelsson and Sköld 2011). The UN Special Rapporteur Paul Hunt has pointed out that it is practically impossible to improve the situation of indigenous peoples if they are not visible through enumeration (Hunt 2007). Ethnicity is not included in population registers in Sweden, Norway and Finland making population estimates difficult (AMAP 2009). Norway, Sweden and Finland have developed systems of population registers that can be linked to social and health data that help to produce accurate, timely demographic data on most vital statistics. However, these registrars are not open for ethnic self-identification which render indigenous people invisible in official statistics. This omission of data is particularly problematic since studies show that the Sami people represent one of few examples of a successful health transition. However, in Norway substantial governmental resources have been assigned to research the health condition of the Sami population and the second wave of SAMINOR studies is currently being evaluated (Eriksen 2015). In Sweden the existing information on Sami health and mortality, that has been exclusively available for research, has not been updated since 2002, and there seems to be disagreement on how to proceed.

It is ironic that Sweden has the most excellent historical sources to perform demographic research in historical contexts. The result is unique population data bases that offers the opportunity to follow each individual in the digitized regions every year from birth to death. This goes from around 1750 to 1900. After that the ethnicity becomes much more difficult to detect, and after 1950 it is even abandoned in Swedish official registers of any kind. Officially there are 20,000 Sami in Sweden, but genealogical data shows that there is more than 50,000 persons with strong Sami kinship.

There is a great need of improving and merging quantitative ethnic information at the individual level in official registers and statistics. This is a prerequisite for the understanding of the present situation, and for a sustainable development of indigenous cultures. The Lancet study concludes that we particularly need to develop Indigenous health data systems in close collaboration with Indigenous peoples, improved Indigenous data identifiers, meaningful Indigenous engagement, strong global networks, further international studies, and development by national governments of targeted policies for Indigenous and tribal health.

11.5 Melting Permafrost and the Release of Infectious Diseases

The last section of this chapter deal with three infectious diseases that were all common in the past, and that are considered as minor threats today. Nevertheless, they all bear a potential risk for the future related to changes in the Arctic. Two hundred years ago infectious diseases dominated the high mortality in the Arctic countries. In Sweden smallpox killed more than 300,000 people between 1750 and 1800, in a country with only 2 million inhabitants. After the introduction of vaccine in the early 1800s smallpox mortality decreased rapidly, and in 1976 the World Health Organization declared the disease as totally eradicated from Earth. This was the first infectious disease that human health prevention actively got rid of (Sköld 1996). 1918–1920 an influenza pandemic swept over the world, killing between 50 and 100 million people. The Arctic, and especially the Indigenous peoples, were heavily struck. In local communities in Alaska the disease killed up to 90% of the entire population (Mamelund et al. 2013). Anthrax is an infection by bacteria that was already mentioned in the bible as a disease of herbivores, it remained a major cause of death for animals all over the planet until the end of the nineteenth century, with occasional, sometimes extensive, contamination of human beings. Untreated the disease has a fatality rate higher than 90% (Schwartz 2009).

What then do these three terrible diseases have in common for the future mortality risks in the Arctic? The answer is the consequences of melting permafrost. Smallpox, influenza and anthrax are relatively resistant to external factors and can survive for long, also in ice. When reports of an anthrax outbreak in Siberia came in 2016, it was stated that the bacteria originated from dead reindeer in 1941 (Nechepurenko 2016; Revitch et al. 2012). Severe epidemics with these diseases would have terrible results, not least since immunity status of the present-day populations is very low, in the case of smallpox non-existing. This is a somewhat neglected consequence of climate change that needs to be highlighted.

References

AMAP (2009) AMAP assessment 2009: human health in the Arctic. Arctic Monitoring and Assessment Programme (AMAP), Oslo

Anderson I et al (2016) 'Indigenous and tribal peoples' health (The Lancet – Lowitja Institute Global Collaboration): a population study'. The Lancet 20 April 2016. doi: http://dx.doi.org/10.1016/S0140-6736(16)00345-7, pp 1–27

Arctic Social Indicators [ASI](2010) (eds) Joan Nymand Larsen, Peter Schweitzer and Andrey Petrov. Nordic Council of Ministers: Nordic Council of Ministers Publishers

Axelsson P, Sköld P (2006) Indigenous populations and vulnerability. Characterizing vulnerability in a Sami context. Annales de Demographie Historique 111(1):115–132

Axelsson P, Sköld P (2011) Introduction. In: Axelsson P, Sköld P (eds) Indigenous peoples and demography. The complex relation between identity and statistics. Berghahn Books, New York, pp 1–14

Baldwin P (1999) Contagion and the state in Europe 1830–1930. Cambridge University Press, Cambridge

Bjerregaard P, Young TK, Curtis T (2003) 35 years of ICCH: evolution or stagnation of circumpolar health research? In: Proceedings of the 12th international congress on circumpolar health, September 10–14, 2003, Nuuk, Greenland. Int J Circumpolar Health 63, suppl. 2; 23–29

Eriksen AMA (2015) Emotional, physical and sexual violence among Sami and non-Sami populations in Norway: the SAMINOR 2 questionnaire study. Scand J Public Health 43:588–596

Hunt P (2007) 'Mission to Sweden: report of the special rapporteur on the right of everyone to the enjoyment of the highest attainable standard of physical and mental health', *Report No. A/HRC/4/28/Add.2*, United Nations general assembly, human rights Council. New York: United Nations

Mamelund S-E, Sattenspiel L, Dimcka J (2013) Influenza-associated mortality during the 1918–1919 influenza pandemic in Alaska and Labrador: a comparison. Soc Sci Hist 37(2):177–229

Nechepurenko I (2016) 'Anthrax outbreak in Russia kills boy, 12, and hospitalizes others'. The New York Times, 2 August 2016

Porter D (1999) Health, civilization and the state: a history of public health from ancient to modern times. Routledge, London

Rehnberg C (1990) *The Organization of Public Health Care. An Economic Analysis of the Swedish Health Care System.* Linköping University, Linköping

Revitch B, Tokarevich N, Parkinson AJ (2012) 'Climate change and zoonotic infections in the Russian Arctic'. Int J Circumpolar Hêlth 71: 18792. http://dx.doi.org/10.3402/ijch.v71i0.18792

Schwartz M (2009) Dr. Jekyll and Mr. Hyde: a short history of anthrax. Mol Asp Med 30(6):347–355

Sköld P (1996) From inoculation to vaccination: smallpox in Sweden in the eighteenth and nineteenth centuries. Popul Stud 50(3):247–262

Sköld P (2015) Perpetual adaption? Challenges for the Sami and reindeer husbandry in Sweden. In: Birgitta E, Larsen JN, Pasche Ø (eds) The new Arctic. Springer International, London, pp 39–55

Sundin J, Willner S (2007) Social change and health in Sweden – 250 years of politics and practice. Swedish National Institute of Public Health, Stockholm

Chapter 12
Uncertainties in Arctic Socio-economic Scenarios

Riina Haavisto, Karoliina Pilli-Sihvola, and Atte Harjanne

Abstract Scenarios are neither predictions nor forecasts, but explore a range of possible futures. Socio-economic scenarios enable the consideration of different uncertainties related to the future and may improve decision-making by enabling the development and analysis of robust decisions. The development of socio-economic scenarios in the Arctic has been a fairly popular topic for scenario analyses. This study reviews ten selected socio-economic scenarios developed for the Arctic region that differ in structure and geographic focus. The analysis shows that the key uncertainties are fairly similar across the different scenarios. The key uncertainties are mainly related to governance or management and natural resources, but recently the uncertainty and importance of political factors have risen. Climate change is included in all scenarios, but its contribution to the future development of the region and its perceived uncertainty varies depending on the scenario.

12.1 Introduction

The Arctic includes regions, countries and communities with different histories, cultures, political environments and economies. Some of these follow the national borders; others do not. However, some characteristics are common across the region: the Arctic is remote, peripheral and sparsely populated, with challenges regarding accessibility and connectivity, its climate is challenging, it has limited socio-economic resources with constraints on human capital, and a proportionally high dependence on the public and primary (extractive) sectors (Stępień 2016). All these characterize the region, and will also characterize, to different extents, its future.

R. Haavisto (✉) • K. Pilli-Sihvola • A. Harjanne
Finnish Meteorological Institute, P.O Box 503, Helsinki FI-00101, Finland
e-mail: riina.haavisto@fmi.fi

© The Author(s) 2017

K. Latola, H. Savela (eds.), *The Interconnected Arctic — UArctic Congress 2016*,
Springer Polar Sciences, DOI 10.1007/978-3-319-57532-2_12

Due to inter-linked global and local developments taking place at the moment, the Arctic and in its sub-regions will witness changes that are difficult to evaluate. In addition to the characteristics of the Arctic, there are several other uncertainty factors that shape its future: for instance globalization, technological development, global economic situation, climate change, climate policy and its implementation, resource prices and demand, and geopolitical situation are all global developments that will have regional and local impacts (Haavisto et al. 2016). Decision-making in the Arctic is in the interface of all these uncertainties. In the face of this uncertainty and the accelerating change witnessed in the Arctic, decision-making is facing increasing challenges, and the robustness of decisions under different futures is key for successful decision-making.

Futures studies address the long-term developments and alternative futures with different approaches, of which scenario development and scenario planning are two examples. Scenario development is a process of creating future stories, while scenario planning is a more comprehensive foresight study methodology that includes scenario development as one part of the analysis. According to Bishop et al. (2007, p.8) *"– a scenario is a product that describes some possible future state and/or that tells the story about how such a state might come about."* Socio-economic scenarios present possible futures and address social and economic change, governance structures, social values and technological change (McCarthy et al. 2001; Shackley and Deanwood 2003; van't Klooster et al. 2011).

In this chapter, the importance of understanding different socio-economic futures in the Arctic context and the potential of socio-economic scenarios to support decision-making are discussed. This is achieved by reviewing a selection of published Arctic socio-economic scenarios and analysing how they have been developed and how they address the key uncertainties regarding the future of the Arctic. A specific focus is put on the role of climate change as an uncertainty factor in the region, because the Arctic is considered to be a climate change hotspot (Diffenbaugh and Giorgi 2012), and climate change is projected to increase the average temperatures in the Arctic substantially more than the global temperatures (Collins et al. 2013). This chapter draws upon existing literature and research, but is motivated by the TWASE- research project (Towards better tailored Weather and marine forecasts in the Arctic to serve Sustainable Economic activities and infrastructure), funded by the Academy of Finland.

12.2   Towards Better Informed Decision-Making in the Arctic

12.2.1   Socio-economic Scenarios

Scenarios have a long history in business planning, and a great deal of the methodological literature on scenarios is based on scenario exercises in business environments (e.g. Schwartz 1991). Nowadays, scenario development is also used for instance in local community management (e.g. van Oort et al. 2015) and in public policy (e.g. van Asselt et al. 2010), where scenario development is a tool to address inherent uncertainty and complexity in environmental decision-making (Carter and White 2012). Despite their potential value, scenarios are not fully utilized in the mainstream decision-making (Carter and White 2012). Socio-economic scenarios, in particular those allowing for uncertainty considerations, can contribute to decreasing decision failures (Berkhout et al. 2002; Chermack 2004). To avoid decision failures, participatory scenario development methods, where the organizations using the socio-economic scenarios developed are involved in their development, should be used. They increase the effectiveness of the scenarios (Berkhout et al. 2002; Chermack 2004; van Drunen et al. 2011), enable policy and organizational learning (Berkhout et al. 2002) and stakeholder engagement (Shackley and Deanwood 2003).

Berkhout et al. (2002) define two major approaches of socio-economic scenario development: the normative and the exploratory approaches. Normative scenario approach reminds of objective-based planning, in which the future is built on positive or negative visions and pathways to the desired end point. Exploratory scenario approach builds on alternative socio-economic conditions, constructing plausible representations of the future and supporting establishment of robust strategies. In comparison to the two approaches in Berkhout et al. (2002), Börjeson et al. (2006) categorize scenarios into predictive, explorative and normative scenarios, predictive scenarios trying to foresee what will or is expected to happen.

Socio-economic scenarios can be used to test and develop robust policies, plans and investments in two ways: (1) by drafting a plan, developing the scenarios, and testing the plan within the scenarios, one ends up with a robust plan; or (2) by developing the scenarios, which inspire plan elements, and testing elements across scenarios, one ends up having a robust plan (Palazzo et al. 2016). Scenarios can influence strategic policy, help the development of forward plans, and assess the potential performance of strategies, policies and actions under different futures. They can also raise awareness of possible factors that drive change and stimulate the development of more robust future oriented decisions and actions. (Carter and White 2012). In addition, they help to open up the uncertainties in the decision-making context (Shackley and Deanwood 2003).

12.2.2   *Socio-economic Scenarios for the Arctic*

The literature on the Arctic future is miscellaneous and consists of various methodological approaches. Arbo et al. (2013) review approximately 50 Arctic futures
studies and identify six methodological approaches in studies dealing with socio-
economic and political aspects: implicit futures, visions, high probability scenarios,
qualitatively different scenarios, simulations, and games. These 50 studies include
all kinds of forward-looking and policy oriented studies, of which not all present
holistic socio-economic scenarios. Yet, socio-economic scenarios may be developed
with any of the above approaches. Implicit futures are seen as a continuation of the
past, leaving little room for uncertainty consideration, and visions tend to be of
predictive nature and have only one future vision. Simulation and game approaches
tend to be modelling exercises including a set of (uncertain) assumptions of socio-
economic change and they may both create socio-economic scenarios and include
them as input to the models.

To complement the study by Arbo et al. (2013) and to identify the recent developments in the field, a selection of different Arctic socio-economic scenarios creating multiple possible futures were identified. The content characteristics of different
Arctic scenarios were of interest; the factors and uncertainties identified within
them, and how they addressed climate change. The starting point for the sampling
was the literature list provided by Arbo et al. (2013).[1] Studies chosen were such
studies that allow reasonable consideration of uncertainties of possible development
trends or futures, and that also create a clear set off different future visions or narratives – not just a single story. However, simulations and games were left out from
the analysis. Furthermore, the list of Arbo et al. (2013) was complemented with
recent studies found by various Google Scholar and Web of Science searches, and
also with the socio-economic scenarios developed by the authors in the TWASE-
project (Haavisto et al. 2016). The following criteria was applied to choose the
items added to the list of Arbo et al. (2013):

1. The language of the study is English and the study is available in an electronic
   format (article, report or website).
2. The geographic focus of the study is on the Arctic as a whole or a more specific
   Arctic area.
3. The study describes the methodology for the socio-economic scenario development and describes the resulting scenarios.
4. The study deals with decision-making or at least has strong connection to it
   (implying that impact studies were left out).

It was acknowledged that this sample does not include all relevant studies
addressing the future of the Arctic; yet, it was believed to give a good overview of
the methods that have been used to create the Arctic socio-economic scenarios relevant for holistic decision-making. The sample also presents qualitatively different

---

[1] http://site.uit.no/arcticfutures/arctic-futures/7-references/

scenarios for the Arctic or its sub-regions, which implies that the relevant uncertainties are considered.

The final sample consisted of 10 studies, listed below in the ascending order of publication year. In addition to the reference, the list consists of the geographic focus, time horizon, decision-making context and the purpose of the scenarios.

- Arbo et al. (2007): the Barents Sea (Finnmark County and Murmansk Oblast); 2030; oil and gas businesses; to discuss the consequences of oil and gas development;
- Brigham (2007): the Arctic; 2040; Arctic development in general; to provoke thinking of the Arctic future;
- AMSA (2009) and GBN (2008): Arctic marine areas; 2050; actors related to marine navigation; to consider the long-term social, technological, economic, environmental and political impacts on Arctic Marine Navigation and providing material for decision-making;
- Cavalieri et al. (2010): the Arctic; 2030; European Union's environmental footprint and policy in the Arctic; to aid the discussion and inform long-term policy development to decrease EU's Arctic footprint;
- Loe et al. (2014): the Arctic; 2020; businesses in the petroleum, mining and seafood sectors, and regional and transit shipping; to enable businesses to identify opportunities and challenges;
- Wesche and Armitage (2014): Slave River Delta, Northwest Territories, Canada; 2030; community level; to understand community vulnerability and adaptation responses to climate change;
- Van Oort et al. (2015): Barents region (Kirovsk in Murmansk Oblast and Bodø in Nordland); 30–50 years ahead; county and municipal levels; to aid climate change adaptation related decision-making;
- Nilsson et al. (2015): Barents Region: Pajala, Sweden; 30–50 years ahead; county and municipal levels; to aid climate change adaptation related decision-making;
- Beach and Clark (2015): Southwest Yukon, Canada; 2032; local and regional wildlife managers; to understand community vulnerability and adaptation responses to climate change;
- Haavisto et al. (2016): Eurasian Arctic; 2040; weather and marine services; to assess the need for and development of weather and marine services in shipping, tourism and resource extraction.

The scenarios were developed following two different lines of technique categories, either based on judgmental choices (n = 4) or based on thinking of the dimensions of uncertainty (n = 6). Judgment refers to not having specific methodological support, whereas dimensions of uncertainty techniques do have support of for example scenario-axes techniques (Bishop et al. 2007). Furthermore, taking into account the uncertainties in development trends is part of the latter technique and included in the whole scenario development processes. The studies addressing uncertainties with dimensions of uncertainty techniques include AMSA (2009) and GBN (2008), Cavalieri et al. (2010), van Oort et al. (2015), Nilsson et al. (2015),

Beach and Clark (2015), and Haavisto et al. (2016). Also three judgement based studies, (Arbo et al. 2007; Brigham 2007, Wesche and Armitage 2014) expressed that there are uncertainties related to the development trends but in the final scenarios (narratives) these might not be addressed. Loe et al. (2014) did not discuss uncertainties.

12.2.3 Key Uncertainties and Climate Change in the Scenarios

All the studies identified and addressed multiple possible factors influencing the future of the Arctic, but not all identify key uncertainties. Key uncertainties are those factors that have high impact on the future development but which occurrence is highly uncertain. From the many possible factors, key uncertainties were identified and narrowed down from one to four in AMSA (2009) and GBN (2008), Cavalieri et al. (2010), Wesche and Armitage (2014), Beach and Clark (2015) and Haavisto et al. (2016). This was mainly done with the scenario axes techniques. For example in Haavisto et al. (2016), the key uncertainties were identified by anonymous web-based voting which resulted in the heat map shown in Fig. 12.1. This, however, was only a starting point for the discussion on the final scenario axes,

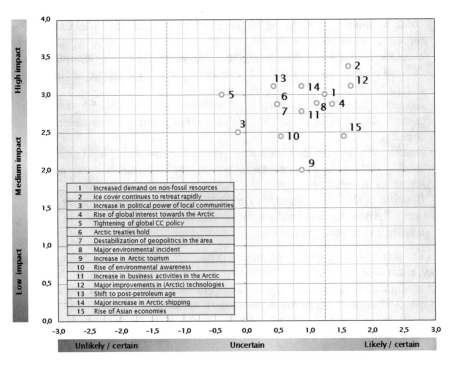

Fig. 12.1 Heat map representing the likelihood/certainty and impact of chosen development trends in Haavisto et al. (2016)

because the resulting key uncertainties were thematically far apart from each other. The key uncertainties (or axes) of the above studies include governance, resources and trade, EU economic growth, EU resource efficiency, climate change in the Arctic, management of environmental pressures in the Arctic, resource development, the human factor, land use, changing ecological-social interactions (which includes climate change), perception of the Arctic as open or closed, initiator of actions being private or public, and dirty or clean environment.

In van Oort et al. (2015) and Nilsson et al. (2015) all factors (16 in Bodø, 10 Kirovsk and 18 in Pajala) were ranked based on their importance and uncertainty but *key* uncertainties have not been indicated – the chosen uncertainties are discussed in the scenario narratives. However, here the rankings were analysed in order to find which factors could be considered to have the highest importance and highest uncertainty. No such factor stands out, except for 'international security' in Pajala. Compared to 'international security', 'climate impacts' are considered less uncertain and important, but 'climate change' is considered a lot less uncertain but a bit more important than 'international security'. In Bodø, the closest factors to key uncertainty were 'global economy' and 'local politics', which have highest uncertainty but a rather low importance, 'climate change' and 'demography', which have a lower uncertainty but higher importance compared to the previous, and 'energy/ petroleum', which has the highest importance but the lowest uncertainty. In Kirovsk, 'climate change' was the most uncertain factor, but had a rather low importance. 'Mineral resource market' and 'foreign policy' were slightly less uncertain but more important and environmental conditions and technological development had the least uncertainty but the highest importance.

In Arbo et al. (2007), the main factor driving the future development of the Arctic was hunt for oil and gas resources, but it was assumed as a certain trend and only uncertainties that may have impact to it were mentioned. Loe et al. (2014) and Brigham (2007) identify key factors but did not address their uncertainties.

Climate change was taken into account in all of the studies but with varying point of views. In Cavalieri et al. (2010), Brigham (2007) and Wesche and Armitage (2014), climate change was appointed as critical uncertainty or key factor, and in Beach and Clark (2015) it was included in critical uncertainty cluster named as "changing ecological-social interactions". In van Oort et al. (2015) and Nilsson et al. (2015), climate change was ranked based on its perceived importance and uncertainty by stakeholders, the ranking depending on the region in question. Climate change received the highest uncertainty and sixth importance ranking in Kirovsk, Russia and "climate change + impacts" fourth uncertainty and second importance ranking in Bodø, Norway (van Oort et al. 2015), and fifth uncertainty and the highest importance ranking in Pajala, Sweden (Nilsson et al. 2015). In Pajala, climate impacts were ranked separately from climate change, and it received second highest uncertainty and eleventh highest importance ranking. In contrary, in the scenarios in Haavisto et al. (2016), climate change was considered to be a high impact but rather certain development trend. Also in AMSA (2009) and GBN (2008) scenarios, climate change was assumed to continue with rather high certainty. In Arbo et al. (2007), climate change was mentioned but not underlined and in Loe et al. (2014), climate change was considered as one of the nine clusters of factors influencing the development but mainly just for the sake of mentioning.

12.3 Discussion

The scenarios in the selected studies were mostly developed through participatory processes; only Arbo et al. (2007) and Brigham (2007) had not involved stakeholders. This is partly due to the strict criteria used in selecting the studies and due to the attempt here to select scenarios oriented toward decision-making. Furthermore, presumably the benefits of stakeholder participation in scenario development (Berkhout et al. 2002; Shackley and Deanwood 2003; Chermack 2004; van Drunen et al. 2011) have motivated the use of participatory methods in the more recent studies. Because of the small sample size and the extensive criteria used to choose it, heavy conclusions cannot be drawn, but those studies that involve stakeholders tend to take uncertainties into account through analytical exercises, not only by relying on single judgement of the uncertainty. Since the review by Arbo et al. (2013), four out of six studies have applied dimensions of uncertainty techniques (Bishop et al. 2007) to develop socio-economic scenarios. The considered uncertainties are case-dependent but similar uncertainties are considered in all of the sample studies. The key uncertainties standing out relate to governance or management and natural resource use. This supports the rather traditional view of the Arctic. However, it seems that uncertainty and importance of political factors are on the rise and already being addressed with higher importance and uncertainty in the more recent studies, published since 2015.

One interesting remark is that the uncertainty related to climate change is seen rather differently across the selected studies. While some consider progressing climate change as a rather certain development, some find it highly uncertain. This may depend on the perceptions of the participants (or the authors) in the scenario development process and also how climate change is actually defined and understood. To some it might mean the impacts of climate change and not the physical phenomenon itself. According to Arbo et al. (2013), the connection between physical climate change and socio-economic scenario development is still lacking in Arctic future studies. Perhaps this is still the case.

Even though the use of socio-economic scenarios in decision-making may decrease decision failures, understanding the uncertainties in scenario definitions and scenario construction, and communicating it to stakeholders is important and requires more research efforts (Mahmoud et al. 2009). To further enhance the use of socio-economic scenarios, research should demonstrate the benefits of scenario approaches in practice (Carter and White 2012).

To be prepared for future changes, to adapt to climate change, and to decrease the vulnerability of societies, the chosen policies and policy measures should be robust and flexible enough so that resources are used as sensibly as possible regardless of the future. The development process of the scenarios and the final socio-economic scenarios have the potential to inform decision-making. Still, further research should address how these kind of scenarios are actually utilized in decision-making and do the final scenarios ever find their way to decision-makers who have not been part of the development process. In theory, socio-economic scenarios have great potential in decision-making but still more research is needed to understand their actual use (Arbo et al. 2013) and benefits (Carter and White 2012).

12.4   Conclusions

The Arctic region is projected to face many changes in the future due to both physical and socio-economic factors. The future development is still far from certain or clear. Scenarios provide a tool to address this uncertainty and strive for robust decisions, and the Arctic has been a popular topic of different scenario analyses. In this review of ten published scenario sets, the presented scenario key uncertainties were typically related to governance or management and natural resources but in more recent scenarios political factors were also brought up. Despite the already witnessed and projected changes, climate change was only considered as one factor among others and its uncertainty differed from fairly certain to uncertain. This may be a particular characteristic of broad socio-economic scenarios or the use of participatory methods since other factors may be more pressing for the stakeholders involved in the scenario development process. There seems to be certain convergence in the themes within Arctic scenarios. Exploring the reasons for this and studying if and how separate scenario development processes affect each other would be interesting topics for future research.

References

AMSA (2009) Arctic marine shipping assessment 2009 report. Arctic Council and Protection of the Arctic Marine Environment (PAME). http://hdl.handle.net/11374/54. Accessed 2 Feb 2017

Arbo P, Didyk V, Hersoug B, Nilssen IB, Nygaard V, Riabova L, Sand JY, Østbye S (2007) Petrodevelopment 2030: Socio-economic consequences of an extensive oil and gas development in the Barents Sea. University of Tromsø. http://hdl.handle.net/10037/1242. Accessed 2 Feb 2017

Arbo P, Iversen A, Knol M, Ringholm T, Sander G (2013) Arctic futures: conceptualizations and images of a changing Arctic. Polar Geogr 36:163–182. doi:10.1080/1088937X.2012.724462

Beach DM, Clark DA (2015) Scenario planning during rapid ecological change: lessons and perspectives from workshops with southwest Yukon wildlife managers. Ecol Soc. doi:10.5751/ES-07379-200161

Berkhout F, Hertin J, Jordan A (2002) Socio-economic futures in climate change impact assessment: using scenarios as "learning machines.". Glob Environ Chang 12:83–95. doi:10.1016/S0959-3780(02)00006-7

Bishop P, Hines A, Collins T (2007) The current state of scenario development: an overview of techniques. Foresight 9:5–25. doi:10.1108/14636680710727516

Börjeson L, Höjer M, Dreborg K-H, Ekvall T, Finnveden G (2006) Scenario types and techniques: towards a user's guide. Futures 38:723–739. doi:10.1016/j.futures.2005.12.002

Brigham LW (2007) Thinking about the Arctic's Future. Futur Sept–Oct 27–34

Carter JG, White I (2012) Environmental planning and management in an age of uncertainty: the case of the water framework directive. J Environ Manag 113:228–236. doi:10.1016/j.jenvman.2012.05.034

Cavalieri, S, McGlynn E, Stoessel S, Stuke F, Bruckner M, Polzin C, Koivurova T, Sellheim N, Stępień A, Hossain K, Duyck S, Nilsson AE (2010) EU Arctic footprint and policy assessment. Ecologic Institute, Berlin. http://arctic-footprint.eu/node/10. Accessed 2 Feb 2017

Chermack TJ (2004) Improving decision-making with scenario planning. Futures 36:295–309. doi:10.1016/S0016-3287(03)00156-3

Collins M, Knutti R, Arblaster J, Dufresne J-L, Fichefet T, Friedlingstein P, Gao X, Gutowski W et al (2013) Chapter 12 – long-term climate change: projections, commitments and irreversibility. In: Climate change 2013: the physical science basis. IPCC working group I contribution to AR5. Cambridge University Press, Cambridge, UK/New York

Diffenbaugh NS, Giorgi F (2012) Climate change hotspots in the CMIP5 global climate model ensemble. Clim Chang 114:813–822. doi:10.1007/s10584-012-0570-x

GBN (Global Business Network) (2008) The future of Arctic marine navigation in mid-century. Scenario Narratives Report. Protection of the Arctic Marine Environment (PAME). http://hdl.handle.net/11374/838. Accessed 2 Feb 2017

Haavisto R, Pilli-Sihvola K, Harjanne A, Perrels A (2016) Socio-economic scenarios for the Eurasian Arctic by 2040. FMI reports 2016:1. Finnish Meteorological Institute. http://hdl.handle.net/10138/160254. Accessed 2 Feb 2017

Loe JS, Fjærtoft DB, Swanson P, Jakobsen EW (2014) Arctic business scenarios 2020. Menon Business Economics. http://www.menon.no/publication/arctic-business-scenarios-2020/. Accessed 2 Feb 2017

Mahmoud M, Liu Y, Hartmann H, Stewart S, Wagener T, Semmens D, Stewart R, Gupta H, Dominguez D, Dominguez F, Hulse D, Letcher R, Rashleigh B, Smith C, Street R, Ticehurst J, Twery M, van Delden H, Waldick R, White D, Winter L (2009) A formal framework for scenario development in support of environmental decision-making. Environ Model Softw 24:798–808. doi:10.1016/j.envsoft.2008.11.010

McCarthy J, Canziani OF, Leary NA, Dokken DJ, White KS (2001) Climate change 2001: impacts, adaptation, and vulnerability. Contribution of working group II to the third assessment report of the intergovernmental panel on climate change. Cambridge University Press, Cambridge

Nilsson AE, Carlsen H, van der Watt L-M (2015) Uncertain futures: the changing global context of the European Arctic: report of a scenario-building workshop in Pajala, Sweden. SEI working paper 2015–12. Stockholm Environment Institute. https://www.sei-international.org/publications?pid=2833. Accessed 2 Feb 2017

Palazzo A, Rutting L, Zougmoré R, Vervoort JM, Havlik P, Jalloh A, Aubee E, Helfgott AES, Mason-D'Croz D, Islam S, others (2016) The future of food security, environments and livelihoods in Western Africa: Four socio-economic scenarios. CCAFS working paper no. 130. CGIAR Research Program on Climate Change, Agriculture and Food Security (CCAFS). http://hdl.handle.net/10568/73375. Accessed 2 Feb 2017

Schwartz P (1991) The art of the long view: planning for the future in an uncertain world. Doubleday/Currency, New York

Shackley S, Deanwood R (2003) Constructing social futures for climate-change impacts and response studies: building qualitative and quantitative scenarios with the participation of stakeholders. Clim Res 24:71–90

Stępień A (2016) Other futures for arctic economies? Searching for alternatives to resource extraction. ArCticle 4/2016. Arctic Centre. http://lauda.ulapland.fi/handle/10024/62539. Accessed 2 Feb 2017

van Asselt MB, van't Klooster SA, van Notten PW, Smits LA (2010) Foresight in action: developing policy-oriented scenarios. Routledge, Washington, DC van Drunen MA, van't Klooster SA, Berkhout F (2011) Bounding the future: the use of scenarios in assessing climate change impacts. Futures 43:488–496. doi:10.1016/j.futures.2011.01.001

van Oort B, Bjørkan M, Klyuchnikova EM (2015) Future narratives for two locations in the Barents region. CICERO Report 2015:06. CICERO Center for International Climate and Environmental Research. http://hdl.handle.net/11250/2367371. Accessed 2 Feb 2017

van't Klooster SA, van Druunen MA, Koomen E (2011) Socio-economic scenarios in climate adaptation studies. In: Climate adaptation and flood risk in coastal cities. Earthscan, London/ New York, pp 27–51

Wesche SD, Armitage DR (2014) Using qualitative scenarios to understand regional environmental change in the Canadian north. Reg Environ Chang 14:1095–1108. doi:10.1007/ s10113-013-0537-0

Chapter 13
Importance of Consideration of Climate Change at Managing Fish Stocks: A Case of Northern Russian Fisheries

Dmitry Lajus, Daria Stogova, and Julia Lajus

Abstract Effect of climate change on the populations of commercial fish is widely recognized. However, this recognition is currently insufficient and climate parameters are not incorporated into fishery forecasting models. Major fisheries of northern Russia targeting Alaska pollock, Pacific salmon in the North Pacific, and Atlantic cod in the Barents Sea are now in a good shape and showing record catches. This review discusses how climate change should be taken into account in the management of northern fish stocks in Russia. Given that climate conditions are currently favorable for these fisheries, it is difficult to assess the effectiveness of management system and predict how it will behave under less favorable climatic situation. Climate change might play a positive role in short-term perspective, but its role may be even negative in long-term perspective because of the possibility that the management system might lose its effectiveness in favorable conditions. To reduce risks for commercial fish stocks, it is necessary to incorporate an ecosystem-based approach in the management. One opportunity for that is provided by the program of ecological certification of Marine Stewardship Council (MSC) which became well established in Russia during the last decade. Without any support from the state, participants of the MSC program educate fishers, fishery managers, and governmental officers towards the use of ecosystem-based approach, specially accounting for the effect of climate change on northern Russian fisheries.

D. Lajus (✉) • D. Stogova
St. Petersburg State University, Saint Petersburg, Russia
e-mail: dlajus@gmail.com

J. Lajus
National Research University Higher School of Economics, Saint Petersburg, Russia

© The Author(s) 2017
K. Latola, H. Savela (eds.), *The Interconnected Arctic — UArctic Congress 2016*,
Springer Polar Sciences, DOI 10.1007/978-3-319-57532-2_13

13.1 Introduction

Since the development of modeling approaches to study dynamics of fish stocks of commercial fisheries[1] in the 1950s, the main aim of fisheries scientists was to determine how much fish can be withdrawn from the stocks without paying much attention to the effect of environmental factors on their population dynamics. Variation in stocks' size therefore was mostly attributed to fishing mortality (Finley 2011). Certainly, effect of weather conditions on recruitment of commercial fish was known since mid-nineteenth century at least, but these fluctuations were considered rather as random effects than as a long-term trends (Cushing 1982). After the book of Cushing provided a convincible set of examples of influence of climate on fish stocks, this influence became widely recognized among scientists, especially since the 1990s when climate change studies in general rapidly developed (Glantz 1992, 1994). By that time, in-depth analysis of several cases of overfishing, in particularly, the Pacific sardine in the 1950s (Lluch-Belda et al. 1989; MacCall 2011), and the Atlanto-Scandian herring in the 1970s (Nakken 2008) has been performed. It occurred that at the time when the overfishing happened, their decline was attributed to problems of management; in fact the overfishing coincided with a period of unfavorable climate conditions, which did not allow management to respond to the situation adequately.

A recognition of climate effects on fisheries, however, is still insufficient. According to Cheung and Pauly (2016, 86), "Of the various ways humans affect marine ecosystems, climate change may be the most insidious and unrecognized. In fact, even if they believe that it is occurring, most people think climate change is going to affect us later, and thus there is no real urgency". Although there were some attempts to formally incorporate climatic parameters into population models, these attempts have so far not been successful (King et al. 2015).

The major Russian commercial fisheries are situated in high latitudes where climate change is very pronounced now, and may thus cause a considerable effect on the stocks. All these fisheries experience a period of high, sometimes record catches, which is often attributed to effective management. A number of these fisheries were awarded with Marine Stewardship Council (MSC) certification showing recognition of management effectiveness internationally. Meantime, these fisheries targeted boreal but not Arctic species, for which warming is usually a favorable factor for their distribution and abundance (Drinkwater et al. 2006; Stige et al. 2010). This indicates that warming alone may play an important role in the current favorable status of the northern Russian fisheries. As the roles of favorable climate conditions and effective management on the good status of the fish stocks are difficult to separate from each other, it is difficult to assess the effectiveness of management system objectively and predict how it will behave under less favorable climate conditions. This review is discussing how climate change should be taken into account in management of northern fish stocks in Russia.

---

[1] According to the Merriam-Webster dictionary (www.merriam-webster.com), term "fishery" has the following meanings: (i) the occupation, industry, or season of taking fish or other sea animals (as sponges, shrimp, or seals): fishing, (ii) a place for catching fish or taking other sea animals, (iii) a fishing establishment, (iv) the legal right to take fish at a particular place or in particular waters, (v) the technology of fishery —usually used in plural.

13.2   Major Fisheries of Northern Russia

The major Russian fisheries comprising altogether more than a half of total fish catch of Russia, accounted almost 4.493 million metric tons (Svedenia 2015). The largest is Alaska Pollock fishery that comprises 36% of the total fish catch in Russia (Svedenia 2015) and is situated in the Far East (Sea of Okhotsk and Bering Sea). Pollock catches were high in the 1980s, but declined afterwards and have grown again during the last decades probably due to favorable climate (Klyashtorin 1998; Johnson 2016). The fishing is conducted from large vessels by using pelagic trawls. High by-catch of juvenile pollock may cause some problems in this fishery (O'Boyle et al. 2013).

Pacific salmon, represented by several species - pink, chum, sockeye, coho and chinook - are caught mostly near the Kamchatka Peninsula and Sakhalin Island in the North Pacific, comprising about 8% of the total Russian catch. Pacific salmon were abundant in the 1930–1940s, after which the stocks declined and grew again since the 1990s, showing a high positive correlation of population size with water temperatures (Beamish and Bouillon 1993; Klyashtorin 1998). Their catches approached record levels in the last decade (Irvine and Fukuwaka 2011). Most salmon are caught by trap nets - passive fishing gear installed on the salmon's spawning migration passages in the sea before they enter rivers for spawning (Fig. 13.1). Because of spawning in rivers, salmon are easily available for the local population, and illegal fishing is rather intensive and difficult to control. Illegal fishing has now reduced greatly in comparison to the 1990s when it was ubiquitous, but it still represents a common practice in the Far East (Simonova and Davydov 2016).

Fig. 13.1 Loading of pink salmon in the Northern Sakhalin Island (Photo by D. Lajus)

Another threat for the wild salmon stocks is represented by hatcheries, that are used to reduce natural mortality by incubating salmon eggs and rearing their juveniles under artificial conditions and then releasing them in the wild. The hatcheries are mostly operated in the southern part of Sakhalin Island and Kuril Islands, although they are not so numerous as in Japan, Canada and USA. Hatchery fish can compete with wild-origin fish, reducing their growth and abundance but are usually less adapted to natural environment, a characteristic that might be especially important in periods of quick climate changes.

Atlantic cod and related species such as haddock and saithe are caught in the Barents and Norwegian seas and comprise almost 11% of total Russian catch. This fishery is managed by the Joint Norwegian-Russian Fisheries Commission. The Arctic cod stock experienced significant fluctuations with high levels in the 1940–1950s, decline in the 1980s, and increase in the recent decades, following climate trends (ICES 2015 and the references herein). The catches reached the record level in 2014 (ICES 2015). The main fishing gear for catching this species is the bottom trawl, that heavily contacts with the sea bottom during fishing operations and may cause serious damage to bottom communities. It also easily results to a by-catch, formed from a comparatively large proportion of non-target species. Research in the Barents Sea has proved a negative effect of bottom trawls on benthic organisms (Denisenko and Zgurovsky 2013). Now codfish fisheries in the Arctic are considered as an example of excellent management (Pristupa et al. 2016), but only a decade ago the same fisheries caused serious concerns due to illegal fishing and overcapacity of fishing fleet (Kalentchenko et al. 2005).

13.3 Risks for Fish Stocks Associated with Climate Change

The analysis of the historical catch dynamics of the forementioned fish stocks shows that they are in accordance with predictions based on analysis of fisheries worldwide (Cheung et al. 2008, 2009; Fernandes et al. 2013; Cheung and Pauly 2016), i.e. they experience a growth in the warmer periods and decline during the colder ones. Another conclusion that can be drawn from these descriptions is that all these fisheries have management problems imposing risks to their sustainability, such as illegal fishing, destruction of bottom communities, and by-catch of juvenile individuals or non-target species. These problems may not play important role in a period of favorable climate conditions, but in a situation of unfavorable climate conditions they may increase the negative role of climate change, causing a serious problem for the sustainability of fish stocks.

Even if climate conditions will stay favorable, the biotic situation in the ecosystems will eventually change -mostly due to changes in populations of forage organisms and predators, and appearance of introduced species, which is predicted to be quite intensive in high latitudes (Cheung et al. 2009). These factors may cause a change of dominant species and thus require an adequate response of management system. This is possible only if all available information about populations of commercial species and their biotic and abiotic interactions will be taken into consideration. Such approach is called "ecosystem-based management". This approach was conceptually developed in the 1990s (Slocombe 1993; McLeod and Leslie 2009), but until now it has not been very common to apply it in practice because of insufficient knowledge about the patterns and nature of processes in the marine ecosystems. In the case of Russian fisheries, climate change may play a positive role in short-term perspective, but its role may be counterwise in a long-term perspective because of the possibility that the management system may lose its effectiveness under the current favorable conditions.

13.4   How to Reduce Risks for Fisheries?

In order to introduce ecosystem-based approach to fishery management in long-term perspective, it is necessary to modify the higher education both in specialized and general universities by introducing relevant courses into the curriculum. At the same time, there are also approaches that do not rely on governmental funding. In particularly, environmental non-governmental organizations, such as the World Wildlife Fund, actively operating in Russia, traditionally play active role in the education of general public. The understanding of the importance of the fisheries sustainability is too weak now in the society, and special effort is required to improve that. Public education towards consuming sustainable seafood began only few years ago by publication of the first seafood consumer guide in Russian (Lajus et al. 2010). Other publications on sustainable fisheries were devoted to principles of ecological certification (Spiridonov and Zgurovsky 2003), illegal fishing of Kamchatka salmon (Dronova and Spiridonov 2008), comparison of consequences of long-line and bottom trawl fishing (Grekov and Pavlenko 2011), critical analysis of Russian fisheries against provisions of Code of conduct of responsible fisheries (Zgurovsky et al. 2013) and analysis of fisheries-related threats to the Arctic ecosystem (Bokhanov et al. 2013).

In short-term perspective, one approach is the introduction of ecological certifications. Among a number of existing certification systems, the most demanded is a voluntary certification according to the standards of Marine Stewardship Council (MSC). MSC develops standards for certification, and the certification itself is performed by an independent company. Certified fisheries obtain marked advantages. MSC program is quite well established in Europe and North America, but comparatively new for Russia with its very different civil and expert structures. By now, in total 15 fisheries are certified, and among them are the fisheries described in this chapter. Only fisheries that deal with export production participate in the program, because of absence of demand on certified fish on the Russian market.

Participation in the MSC program requires all participants – company managers, governmental authorities, researchers, independent experts –very profound understanding of its standards, consisting of three principles: maintenance of healthy status of target species, limited ecosystem effect of fishing operations, and effective management. During the almost decade of operation of the MSC program in Russia, there has been several publications about MSC principles and process. Also, several seminars have been organized by environmental NGOs such as WWF, MSC, Wild Salmon Center (WSC), Ocean Outcomes (O2) and fisheries clients in Moscow, Murmansk and the Far East. The participants of these seminars have been fishers, vessel owners and principal crew, governmental officers, independent experts. As a consequence of these activities, there is now a number of experts in Russia who are familiar with the process and principles of the MSC program. There is as well a Russian MSC representative, and a Russian certification body. Therefore, the MSC certification currently represents a well-established process that may serve as a framework for spreading the principles of ecosystem-based management to fisheries. Climate change problematics certainly plays a key role in the introduction of ecosystem-based approach to northern Russian fisheries, because climate change is most pronounced in high latitudes and because commercial species are in general biologically very sensitive to such changes. Climate effects are not linear and thus very difficult to forecast, and therefore only cooperative efforts of both managers, researchers and fishers allow to adequately response to their challenges.

Acknowledgements The study was conducted in the framework of the international project "New Governance for Sustainable Development in the European Arctic" funded by The Swedish Foundation for Strategic Environmental Research (MISTRA).

References

Beamish RJ, Bouillon DR (1993) Pacific salmon production trends in relation to climate. Can J Fish Aquat Sci 50:1002–1016

Bokhanov DV, Lajus DL, Moiseev AR, Sokolov KM (2013) Otsenka ugroz morskoi ekosisteme Arktiki, sviazannykh s promyshlennym rybolovstvom, na primere Barentseva moria. WWF Russia, Moscow. https://www.wwf.ru/data/publ/marine/arctic-ecosystems_fish.pdf

Cheung WWL, Pauly D (2016) Global-scale responses and vulnerability of marine species and fisheries to climate change. In: Pauly D, Zeller D (eds) Global atlas of marine fisheries: a critical appraisal of catches and ecosystem impacts. Island Press, Washington, DC, pp 86–97

Cheung WWL, Close C, Lam VWY, Watson R, Pauly D (2008) Application of macroecological theory to predict effects of climate change on global fisheries potential. Mar Ecol Prog Ser 365:187–193

Cheung WWL, Lam VWY, Sarmiento JL, Kearney K, Watson R, Pauly D (2009) Projecting global marine biodiversity impacts under climate change scenarios. Fish Fish 10(3):235–251

Cushing DH (1982) Climate and fisheries. Academic Press, London/New York

Denisenko SG, Zgurovsky KA (2013) Vozdeistvie tralovogo promysla na donnye ekosistemy Barentseva moria i vozmozhnosti snizheniia urovnia negativnykh posledstvii. WWF-Russia, Murmansk. https://www.wwf.ru/data/publ/bentos-i-traly_doklad-wwf.pdf

Drinkwater KF (2006) The regime shift of the 1920s and 1930s in the North Atlantic. Prog Oceanogr 68:134–151

Dronova N, Spiridonov V (2008) Illegal, unreported, and unregulated Pacific Salmon fishing in Kamchatka. WWF Russia/TRAFFIC Europe. http://www.wwf.ru/resources/publ/book/eng/313

Fernandes JA, Cheung WWL, Jennings S et al (2013) Modelling the effects of climate change on the distribution and production of marine fishes: accounting for trophic interactions in a dynamic bioclimate envelope model. Glob Chang Biol 19:2596–2607

Finley C (2011) All the fish in the sea: maximum sustainable yield and failure of fisheries management. The University of Chicago Press, Chicago

Glantz MH (ed) (1992) Climate variability, climate change, and fisheries. Cambridge University Press, Cambridge

Glantz MH (1994) The impacts of climate on fisheries, UNEP Environmental Library 13. UNEP, Nairobi

Grekov AA, Pavlenko AA (2011) A comparison of longline and trawl fishing practices and suggestions for encouraging the sustainable management of fisheries in the Barents Sea. WWF Russia, Moscow/Murmansk. http://wwf.ru/resources/publ/book/eng/456

ICES. Report of the Arctic Fisheries Working Group (AFWG), 23–29 April (2015) ICES CM 2015/ACOM:05, Hamburg http://www.ices.dk/sites/pub/Publication%20Reports/Expert%20 Group%20Report/acom/2015/AFWG/01%20AFWG%20Report%202015.pdf

Irvine JR, Fukuwaka M-A (2011) Pacific salmon abundance trends and climate change. ICES J Mar Sci 68(6):1122–1130. doi:10.1093/icesjms/fsq199

Johnson T (2016) Climate change and Alaska fisheries. University of Alaska Fairbanks, Alaska Sea Grant. http://doi.org/10.4027/ccaf.2016

Kalentchenko M, Nagoda D, Esmark M (2005) Analysis of illegal fishery for cod in the Barents Sea. WWF-Russia. http://www.wwf.se/source.php/1120186/WWF_Russia_IUU%20fishing_barents_2005_august_1.pdf

King JR, McFarlane GA, Punt AE (2015) Shifts in fisheries management: adapting to regime shifts. Phil Trans R Soc B 370:20130277. http://dx.doi.org/10.1098/rstb.2013.0277

Klyashtorin LB (1998) Long-term climate change and main commercial fish production in the Atlantic and Pacific. Fish Res 37:115–125

Lajus DL, Lajus JA, Zgurovsky KA, Spiridonov VA, Chuzhekova TA (2010) A vy znaete, chto pokupaete? Ekologicheskoe rukovodstvo dlia pokupatelei i prodavtsov rybnoi produktsii. WWF-Russia, Moscow. http://wwf.ru/resources/publ/book/390

Lluch-Belda D, Crawford RJM, Kawasaki AD et al (1989) World-wide fluctuations of sardine and anchovy stocks: the regime problem. S Afr J 8:195–205

MacCall AD (2011) The sardine-anchovy puzzle. In: Jackson JBC, Alexander K, Sala E (eds) Shifting baselines: the past and the future of ocean fisheries. Island Press/Center for Resource Economics, Washington, DC, pp 47–57

McLeod KL, Leslie HM (eds) (2009) Ecosystem-based management for the oceans. Island Press, Washington, DC

Nakken O (2008) Norwegian Spring-Spawning Herring & Northeast Arctic cod : 100 years of Research and management. Tapir Akademisk Forlag, Trondheim

O'Boyle R, Japp D, Payne A, Devitt S (2013) Russian sea of Okhotsk mid-water trawl walleye pollock (*Theragra chalcogramma*) fishery. Public certification report. Vladivostok. https://fisheries.msc.org/en/fisheries/russia-sea-of-okhotsk-pollock/@@assessments

Pristupa AO, Lamers M, Amelung B (2016) Private informational governance in post-soviet waters: implications of the marine Stewardship Council certification in the Russian Barents Sea region. Fish Res 182:128–135. http://dx.doi.org/10.1016/j.fishres.2015.07.006

Simonova VV, Davydov VN (2016) Non-compliance with fishery regulations in Sakhalin Island: contested discourses of illegal fishery. Int J Hum Cult Stud 3(3):232–245

Slocombe DS (1993) Implementing ecosystem-based management. Bioscience 43:612–622

Spiridonov VA, Zgurovsky KA (2003) Ekologicheskaia sertifikatsia morskogo rybolovstva. Informatsia dlia rybakov, korotye ne khotiat, chtoby ikh deti i vnuki ostalis bez ryby. WWF Russia, Apelsin. Vladivostok. http://wwf.ru/resources/publ/book/212

Stige LC, Ottersen G, Dalpadado P, Chan K-S, Hjermann DØ, Lajus DL, Yaragina NA, Stenseth NC (2010) Direct and indirect climate forcing in a multi-species marine system. Proc R Soc B 277:3411–3420

Svedenia ob ulove ryby, dobyche drugikh vodnykh biologicheskikh resursov i proizvodstve rybnoi produkhtsii iz nikh, proizvodstve produktsii tovarnoi akvakultury (tovarnogo rybovodstva) za ianvar – dekabr 2015 g (narastaiushim itogom) (2015) Official website of the Federal Fishery Agency. http://fish.gov.ru/files/documents/otraslevaya_deyatelnost/ekonomika_otrasli/statis-tika_analitika/f407-0_01-12_2015.pdf

Zgurovsky KA, Lajus DL, Moiseev AP et al (2013) Kommentarii ekspertov k kodeksu vedenia otvetstvennogo rybolovstva. WWF Russia, Moscow. https://www.wwf.ru/data/publ/550/kodex-fish_web.pdf

Chapter 14
Preservation of Territories and Traditional Activities of the Northern Indigenous Peoples in the Period of the Arctic Industrial Development

Elena Gladun and Kseniya Ivanova

Abstract In Russia the right to traditional use of lands, biological and other resources such as reindeer pastures, harvesting fauna, fish, non-wood resources of forest including wild plants is declared with due regard to the priorities of indigenous peoples. However, in practice the northern indigenous communities can hardly get an access to their traditional lands. They cannot become owners of hunting lands, fishing areas, cannot obtain long-term licenses for the wildlife use rights, quotes for fishing. Due to many reasons the northern indigenous peoples are not able to compete with major industrial companies. As a consequence, the indigenous peoples do not conduct traditional economic activities, nor do they preserve their traditional lifestyle, values and language. Alongside with guaranteed rules concerning indigenous rights, in the Russian legislation there is a gap in proper regulations of traditional territories use. In the current period of intensive industrial development of the Arctic the legal rules should be revised and supplemented with effective mechanisms of granting and protection of traditional territories and activities of the northern indigenous peoples.

14.1 Introduction

The Russian Federation is a multiethnic society and home to more than 180 peoples. Of these, 40 are legally recognized as "indigenous, small-numbered peoples of the North, Siberia and the Far East". This status is tied to the conditions that a people has no more than 50,000 members, maintains a traditional way of life, inhabits

E. Gladun (✉)
Public and Finance Law Department, Tyumen State University, Tyumen, Russia
e-mail: efgladun@yandex.ru

K. Ivanova
Constitutional and Municipal Law Department, Tyumen State University, Tyumen, Russia
e-mail: k.a.ivanova@utmn.ru

K. Latola, H. Savela (eds.), *The Interconnected Arctic — UArctic Congress 2016*,
Springer Polar Sciences, DOI 10.1007/978-3-319-57532-2_14

certain remote regions of Russia and identifies itself as a distinct ethnic community (International Work Group for Indigenous Affairs IWGIA 2016). The population of indigenous peoples of the North, Siberia and Far East of Russia is 243,982 peoples which makes 0.2 of the total population. In Russia the northern indigenous peoples include the Aleuts, Koryak, Eskimos, Chukchi, Evenks, Yakuts, Yukagirs, Dolgan, Selkup, Nanai, Khanty, Mansi, Nenets, Saami and others (News Agency "Arctic-Info" 2016a, b) . The northern indigenous peoples traditionally inhabit huge territories stretching from the Kola Peninsula in the west to the Bering Strait in the east, which make up about two-thirds of the Russian territory. They inhabit more than 20 federative regions (called "subjects of the Russian Federation"), including the Republic of Sakha (Yakutia), the Kamchatka Territory, the Krasnoyarsk Territory, the Khabarovsk Territory, the Magadan Region, the Murmansk Region, the Chukotka Autonomous Area, the Nenets Autonomous Area, the Khanty-Mansi Autonomous Area -Yugra and the Yamalo-Nenets Autonomous Area (Batyanova et al. 2009). (Fig. 14.1)

It is clear that the northern indigenous peoples have undergone significant changes, which have distanced them from their forefathers in economic, social, cultural, and even anthropometrical respects. However, certain groups of the contemporary indigenous population still preserve both the cultural identity and the economic activities which are considered to determine a traditional lifestyle and pattern of settlement (nomadic or semi-nomadic lifestyle, etc.). Many of them have traditionally been hunters, gatherers, fishermen and reindeer breeders, and these activities still constitute vital parts of their livelihoods (IWGIA 2016).

The features which characterize the northern indigenous peoples are determined by their environment. Their small population size also results from external factors and does not indicate either under-development or inherent population decline. On the contrary, for their specific geographic environment and economy type, a small population size represents an optimal solution (Gumilev and Kurkchi 1989). Their life-support system is closely linked to traditional lands and land use, to the challenging climate and geography conditions – severe weather, limited natural resources, and dispersed settlements. In small groups the indigenous peoples can easily respond to major climatic and environmental changes by altering group sizes, relocating, and being flexible with seasonal cycles in hunting or employment (Park 2008). They preserve the national identity and traditional knowledge, while living in the small villages or separate remote *chums* (living tents made of reindeer skins that are laid over a skeleton of long wooden poles), engaged in traditional activities. Therefore, their number is almost not growing, while the birth rate is high enough (Artyunov 2015). Here it's interesting to mention the results of studies provided by Ljudmila Osipova, the Head of Laboratory of Population Genetics in the Russian Academy of Sciences. In 1990s, conducting the research of the Far North indigenous peoples, she argued that the peoples of the North are "a genetic reserve of the country": for thousands of years in the North they have undergone through the hard natural selection, and the most stable and strong people could only survive (Judina 2010). However the same factors which ensured the high degree of adaptability of northern populations to their extreme living conditions, also made it difficult for

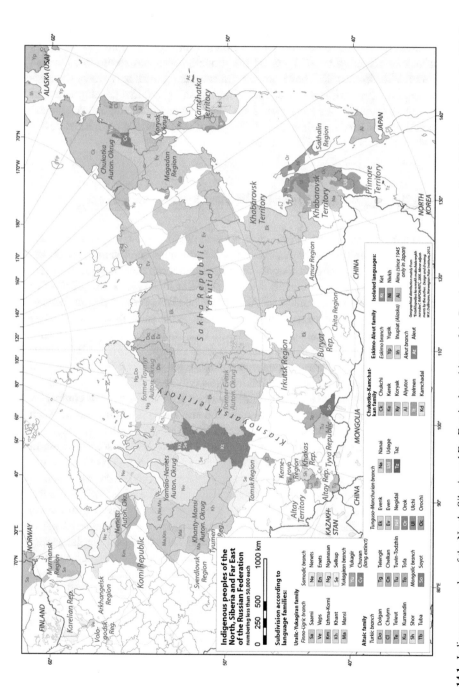

Fig. 14.1 Indigenous peoples of the North, Siberia and Far East in the subjects of the Russian Federation (Goble, Paul. 2016. Numerically Small Indigenous Peoples of the Russian North an Ever Bigger Problem for Moscow. http://upnorth.eu/numerically-small-indigenous-peoples-of-the-russian-north-an-ever-bigger-problem-for-moscow/. Accessed 10 Jan 2017)

them to integrate with other cultures and to adjust to continuing development of their primordial territories.

Notwithstanding cultural and ethnographic differences, indigenous peoples in the Arctic share a common history of assimilation into their various mother states and a lack of recognition of rights on traditionally occupied territories. On this background these peoples have united in working steadfastly towards self-determination. Their primary objective has been to safeguard that development and activity on traditional lands and waters takes into consideration indigenous peoples' traditional way of life, as well as their fair share of the economic benefits of development on these territories (Fløistad 2010).

14.2   Changing Conditions in the Arctic Territories

Traditional use of natural resources – fish, forests, wildlife – is a way of life for the indigenous communities of the Arctic. Indigenous peoples continue to rely on the sustainable use of renewable resources and this dependence puts them at great risk from industrial objects and pollutants that find their way in the period of the intensive economic development. Nowadays the indigenous peoples in Arctic regions have to use alternative ways of their economic development, at the same time they are seeking to balance these emerging opportunities with their traditional lifestyle and values closely connected with the land and wildlife. A good example here is the indigenous peoples of the Yamalo-Nenets Autonomous Area with 40% leading a traditional nomadic life, living right in the forest tundra (News Agency "Arctic-Info". 2016a, b) and sharing their traditional territories with the major Russian oil and gas companies (Gasprom, Rosneft and others) (Gasprom official site 2008). The relations between industrial companies and indigenous peoples today are built on the basis of cooperation agreements and dialogue so that the indigenous peoples receive some short-term benefits of the projects at the stage of arrangement and the operation of oil and gas fields. The public authorities and companies identify the necessary measures to reduce the burden on traditional habitat and traditional economic activities of indigenous peoples of the North. Also the compensation is paid for loss of profit when hunting, fishing, gathering and pastures areas are withdrawn for industrial needs (Bykovskii 2013). On the one hand, this model contributes to improving the quality of life and economic benefits of the indigenous peoples, and on the other it is the path to degradation of peoples accustomed to traditional activities on their lands, who with the arrival of a large number of new people and non-traditional activities have become more vulnerable financially and spiritually (Fig. 14.2).

The economic policy of Russia in the new century focuses on the exploration of mineral resources in the northern Arctic territories and the development of the Arctic energy resources (The Foundations of Russian Federation Policy in the Arctic until 2020 and beyond 2009). Much more than benefits the indigenous peoples of the Russian North face threats from intensive mineral, oil, and gas development, and the resulting conflicts with intensive industrial development model of the

Fig. 14.2 Nenets peoples sharing their lands with modern industrial objects (Foto taken by E. Gladun. 2004)

northern territories have affected all aspects of their life, including social, cultural and spiritual integrity. Yet in April 2005 the 5th Congress of Indigenous Peoples of the North, Siberia and the Far East of the Russian Federation took place in Moscow and gathered more than 300 delegates from 28 subjects of the Russian Federation. The main theme for the Congress was modern social, economic and political processes in the northern territories of Russia and changes in the life of the northern indigenous peoples. Some negative consequences were outlined by indigenous peoples themselves, for example (Gladun and Chebotarev 2015):

– the destruction of the social infrastructure and the public system of medical, cultural, goods, social and transport provision in the places inhabited by the northern indigenous peoples, as a result of which the indigenous peoples involved in reindeer herding and handicrafts, leading a permanent nomadic or semi-nomadic way of life, became completely isolated;
– a deep crisis in the traditional branches of economy, which form the basic life-support of the northern peoples, as a result of ill-considered and swift privatization of the main traditional means of production;
– a decrease of the amount of the indigenous peoples pursuing traditional occupations, as a result of which a general, permanent unemployment is reported, which has led to impoverishment, abrupt increase of morbidity, especially through tuberculosis, and, as a consequence, to a mortality increase and a reduction of the life expectancy for northern indigenous peoples (Gladun and Chebotarev 2015).

In addition, some legal issues were discussed at the Congress – guarantees for the rights of the indigenous peoples and the effective interaction between indigenous communities and organizations with the authorities and industries to implement federal and regional programs of economic and social development of the indigenous peoples (ВСЈУХ.RU 2005). However, significant changes in legislation have not happened afterwards. In fact, many of the achievements of the federal legal regulation made after enactment of 1993 Russian Constitution were annulled by a series of amendments in 2004. The conditions of life and activities of the indigenous peoples of the North in the mid-2000s became significantly different in various regions of Russia. The considerable political, legal and economic efforts to guarantee indigenous rights and to provide support for indigenous communities have been undertaken by regional authorities in such subjects of the Russian Federation as the Republic of Sakha (Yakutia), the Nenets Autonomous Area, the Khanty-Mansi Autonomous Area – Yugra, the Yamalo-Nenets Autonomous Area. In some regions the living conditions and opportunities to maintain traditional activities, to preserve traditional culture and languages continued to deteriorate.

Facing an unprecedented combination of rapid and stressful changes involving environmental forces like climate change, socioeconomic pressures associated with globalization (Arctic Human Development Report (AHDR) 2004; Nuttall 2000) and intensive industrial development (Gladun and Chebotarev 2015) the indigenous peoples in the Arctic have felt a need to safeguard their culture and traditional way of life.

14.3   Legal Regulations on the Indigenous Peoples in the Russian Federation

The existence of legislation on indigenous peoples of the North is an official acknowledgement of their specific legal status by the state. This ensures both the assumptions to preserving of their culture, and promotes a better adaptation of those peoples to present social and economic conditions (Kryazhkov 2013). In the Russian Federation over the last 20 years legislation on indigenous peoples of the North has been formulated as a specific multisectoral development of Russian law. In view of the federal system of government in Russia and related constitutional provisions (arts. 71, 72, 76) (The Constitution of the Russian Federation 2014), this legislation is a two-level, i.e. consists of two blocks of laws and regulations – federal and regional (of so-called "subjects" of the Russian Federation). Federal laws regulate human and civil rights and freedoms of the indigenous peoples, general principles of organizing their traditional territories and activities, the state guarantees for traditional way of life – these and some other issues currently fall in the competence of the federal authorities.[1] The regional regulative level is supplemental, more specific, and remedial. The content of regional indigenous legislation varies to a big

---

[1] Federal legislation on indigenous issues consists of the Concept for the Sustainable Development of Indigenous Peoples of the North, Siberia and the Far East of the Russian Federation (2009), the Federal Law "Guarantees of Rights of Indigenous Peoples in the Russian Federation" (1999), the Federal Law "Territories of Traditional Resource Use" (2001) and some rules and regulations in the specific laws such as the Land Code (2001), the Water Code (2006), the Forestry Code (2006).

extent, but, generally speaking, the regional acts regulate the structure and activities of indigenous communities, formation of territories of traditional resource use, participation of the indigenous peoples in public affairs, traditional economic activities, preservation of aboriginal culture and languages.[2] As mentioned above, the gaps in the federal legislation are sometimes filled in the regional acts.

For example, the Statute of the Yamalo-Nenets Autonomous Area (Ustav (Osnovnoi Zakon) Yamalo-Nenetskogo Avtonomnogo Okruga 1998) guarantees the right of indigenous peoples to participate in the work of regional authorities, local governments according to their national traditions and customs. Public authorities are obliged to take into account the indigenous peoples' opinion when dealing with issues that affect their interests (Устав (Основной закон) Ямало-Ненецкого автономного округа [Charter of the Yamalo-Nenets Autonomous Area] 1998).

The legislation on the indigenous peoples in the Khanty-Mansi Autonomous Area -Yugra is recognized the most developed in the country in the related issues. Thus, the regional authorities provide the guarantees of the indigenous rights and interests by the following means:

– assist for self-government development in accordance with national traditions and customs of indigenous peoples;
– involve the indigenous communities into decision-making process;
– support and finance traditional indigenous activities;
– support indigenous arts, culture and languages;
– preserve and give the priorities for using traditional lands and territories (Устав (Основной закон) Ханты-Мансийского автономного округа – Югры [Charter of the Khanty-Mansi Autonomous Area - Yugra] 1995).

In the Yamalo-Nenets Autonomous Area a new law was enacted in 2016 providing for some measures of social support to the indigenous peoples maintaining traditional lifestyle. These measures include medical support, financial support, and educational support. The most significant achievement of the region is that supportive measures are provided for the indigenous peoples in accordance with their nomadic way of life – new medical centers and schools are organized in the reindeer migration routes and traditional villages of the indigenous communities (Закон Ямало-Ненецкого автономного округа "О гарантиях прав лиц, ведущих традиционный образ жизни коренных малочисленных народов севера в Ямало-Ненецком автономном округе"[On Guarantees of Rights of Peoples Leading Traditional Way of Life in the Yamalo-Nenets Autonomous Area] 2016). Much effort in the region is taken for establishing the dialogue between the indigenous peoples and oil and gas industry (Smorchkova 2015).

---

[2] Examples of regional laws are: "State Support to Indigenous Peoples, to their Communities and the Northern Organizations Involved in Traditional Occupations in the Territory of Yamalo-Nenets Autonomous Area" (2005) The Law "Protection of Traditional Habitat and Traditional Way of Life of Indigenous Peoples of the North in Yamalo-Nenets Autonomous Area" (2006), "Development of Reindeer Herding in the Khanty-Mansi Autonomous Area – Yugra" (2004), "On Reindeer Breeding in the Yamalo-Nenets Autonomous Area" (1998), "Northern Domestic Reindeer Herding in the Republic of Sakha (Yakutia)" (1997) and many others.

In spite of all positive results, the regions are quite restricted by the federal legislation. Both on the federal and regional levels there is a lack of mechanisms to implement the northern indigenous peoples' rights, guaranteed by the Constitution and the federal legislation related to land use, self-government, development of traditional occupations and cultures; there is no well-considered system of regional and branch laws and other normative legislation, which has made it impossible to implement the declared rights. As the result, both federal and regional regulations are fragmentary, there is no system of public authorities providing for protection, guarantees, cooperation, and other forms of preservation of territories and traditional activities of the northern indigenous peoples. That is the reason why despite provided legal regulation there is still a high demand to support northern indigenous peoples, to regard the high vulnerability of their traditional activities, culture and ethnic identity under the conditions of globalization and Arctic industrial development.

We must admit that not only the Russian Federation is facing such challenges. High unemployment, along with health, social, and economic problems, has become a serious issue in other Arctic states (Arctic Climate Impact Assessment (ACIA) 2004). Among the greatest problems for many northern indigenous peoples' are: permanent settlement, relocation, urbanization, climate change as well as the northward advancement of agriculture, introduction of elaborate infrastructure and migration from the South (e.g. fossil fuels extraction, new jobs in public services and tourism) (Arctic Human Development Report (AHDR) 2004; Nuttall 2000, 2002)

Focusing to Russia, the current situation is exacerbated by 'legal stagnation' and a step back from former positions of the state's participation and protection of the territories and traditional activities of the northern indigenous peoples (Kryazhkov 2013). The following examples illustrate the point:

– The lack of any new notable legal acts in this area. During the last decade the main positive achievements were the legal settling of the issue regarding registration of persons, pertaining to indigenous peoples, and the maintaining of the nomadic or semi-nomadic way of life (Federal Law 2011). The Federal law "Territories of Traditional Nature Use of the Indigenous Peoples of the North, Siberia and Far East of the Russian Federation" adopted in 2001 has not been observed and amended for 8 years (Federal Law 2001).
– The repeal in 2004 of the Federal Law "The Basics of the State Regulation of Social and Economic Development of the North of the Russian Federation" (Federal Law 1996). This political decision may be characterized as a denial by the state of the special policy considering the specifics of the northern regions and indigenous peoples living there (Kryazhkov 2013).
– The withdrawal of several provisions related to indigenous peoples from the federal legislation in the period of 2004–2016.[3] As the result, at present, it is no longer possible for indigenous peoples to obtain plots of land for lifetime ownership with hereditary succession and be able to use them free of charge; the allotment of lands for traditional fishing and hunting are provided on the general legal basis.
– The main federal law regulating aboriginal land rights – "On Territories of Traditional Natural Resource Use" – lacks effective mechanisms of legal protec-

---

[3] SZRF: 35 (3607).

tion of traditional territories and activities of the northern indigenous peoples. In its articles the law sets out the legal regime of traditional territories and refers to other federal laws that regulate the land rights and resource-related rights. However the law doesn't set the conditions under which land rights are provided and protected because this is the scope of the federal legislation on land issues. Thus, in Russian legislation there are no norms gramting specific rights to lands and resources for indigenous people who need these for traditional occupations (Gladun 2015).

- By the current federal law fishing and hunting areas are subject to tendering and bidding procedure and there are no exceptions for the indigenous communities inhabiting those territories. As the result, fishing and hunting areas are leased for long-term tenure to fishery and industrial companies (Naikanchina 2010).

Commenting on the present situation Vladimir Kryazhkov, a famous Russian researcher of the indigenous peoples rights and legal regulations,[4] wrote: "The Federal Government does not fulfill its obligation to adopt the necessary regulations; for a long time it rejects requests on organizing traditional territories, focusing on law amendments it does not take particular measures and does not involve practical mechanisms for traditional territories' organization. De facto: the policy of the Federal Government violates the rights of indigenous peoples to traditional resource use and a traditional lifestyle" (Kryazhkov 2008). This opinion is shared by Sergei Kharjuchi, the ex-President of Indigenous Peoples of the North, Siberia and Far East Association. He says: "Up to now there have not been any traditional territories organized on the federal level and the number of regional traditional territories is too small. Moreover, there is only one federal law with a few instruments protecting indigenous rights to traditional lands and lifestyle and this law is not enforced properly because of the inaction of federal authorities" (Отчетный доклад Президента Ассоциации коренных малочисленных народов Севера, Сибири и Дальнего Востока С.Н. Харючи [Annual Report of the President of the Russian Association of Indigenous Peoples of the North, Siberia and Far East S.N. Kharjuchi] 2009). As the result, the northern indigenous peoples in Russia cannot enjoy their rights to traditional land and resource use. The lands used by indigenous peoples may at the same time be used by the oil and gas industry, agriculture industry, landowners. The oil and gas industry and indigenous peoples have been increasingly coming into contact with each other and more exploration and development takes place in lands that indigenous peoples have traditionally occupied.

14.4 Conclusion

To protect indigenous rights to land, natural resources and traditional activities means to set rules and standards for proper regulation of these issues in legislation. As mentioned above, the Russian federal legislation fails to recognize the need for indigenous peoples to use lands freely to maintain their traditional way of life. Legal

---

[4] Vladimir Kryazhkov is a professor of constitutional and administrative law in National Research University "Higher school of Economics".

rules concerning indigenous rights to lands, territories and resources are character-ized as non-compliable with constitutional provisions, incomplete, declarative and they do not imply mechanisms for their enforcement.

To safeguard the legal rights and interests of indigenous peoples especially in the period of the Arctic development, it is essential to create additional mechanisms for the industrial exploitation of traditional territories and nature use areas of indige-nous peoples:

1. To provide and secure gratuitous long-term use of land and traditional natural resources by the northern indigenous peoples, which is important for the preser-vation and development of their traditional way of life (Resolution of the 5th Congress of Indigenous Peoples of the North, Siberia and the Far East of the Russian Federation 2005).
2. To conduct environmental impact assessment in any case of industrial use of lands and natural resources as well as of land acquisition for public purposes in the territories of traditional habitat and occupation of the northern indigenous peoples.
3. To regulate social and economic development of the indigenous peoples on the regional level and eliminate their unemployment through state support of mod-ern development of the traditional livelihoods, like thorough reshaping of rein-deer herding, fishing, sea fishing, gathering of wild plants and handicraft, and the marketing of their products (Resolution of the 5th Congress of Indigenous Peoples of the North, Siberia and the Far East of the Russian Federation 2005).
4. To take into consideration the uniqueness of the traditional way of life and cul-ture of the indigenous peoples when organizing medical services, the education system and other social services (Resolution of the 5th Congress of Indigenous Peoples of the North, Siberia and the Far East of the Russian Federation 2005). There are some good examples of how it might be implemented – nomadic schools and hospitals which have been introduced in the Yamalo-Nenets Autonomous Area in Russia.[5]
5. To provide for representation of indigenous peoples in public authorities so that the indigenous peoples in their residence areas can be represented in electoral committees and can nominate members to legislature and recommend people from their communities to be included in the corresponding party lists (Resolution of the 5th Congress of Indigenous Peoples of the North, Siberia and the Far East of the Russian Federation 2005).

The rights for preserving traditional lifestyle, native language, original culture and transfer of traditional knowledge are inseparable from the basic right of the indige-nous peoples – to use their traditional territories and lands freely. The primary objec-tive of the state is to guarantee this right and to provide opportunities for the indigenous

---

[5] Implementation of the regional project "Nomadic School", aimed at quality improvement in edu-cation and maintenance of traditional lifestyle of indigenous peoples of the North continues in Yamalo-Nenets Autonomous District. http://www.uarctic.org/news/2015/8/nomadic-school-proj-ect-in-yamal. Accessed 20 Dec 2016.

Fig. 14.3 Nenets peoples in the Yamalo-Nenets Autonomous Area (Foto taken by E. Gladun. 2004)

communities. This should be done without paternalism and imposing modern models of cultural development under new economic conditions. The point is that without the right to free use of lands and natural resources, the right of access for using them in compliance with effective legal mechanisms and procedures, there is little sense in constitution guarantees and general international regulations (Fig. 14.3).

Summing up, to ensure the continuous prosperity of the Arctic region Russia needs to follow the main principle stated at the article 22 Rio Declaration on Environment and Development. Indigenous peoples and their communities have a vital role in the Arctic development because of their knowledge and traditional practices. States should recognize and duly support their identity, culture and interests and enable their effective participation in the achievement of sustainable development.

References

Закон Ямало-Ненецкого автономного округа "О гарантиях прав лиц, ведущих традиционный образ жизни коренных малочисленных народов севера в Ямало-Ненецком автономном округе"[On Guarantees of Rights of Peoples Leading Traditional Way of Life in the Yamalo-Nenets Autonomous Area] (2016) No. 1-zao
Arctic Climate Impact Assessment (ACIA) (2004) Impacts of a warming Arctic. Cambridge University Press; United Nations Permanent Forum on Indigenous Issues. 2009. Indigenous peoples, Indigenous voices. Factsheet
Arctic Human Development Report (AHDR) (2004) Stefansson Arctic Institute, Akureyri
Artyunov S (2015) Почему исчезают северные народы? [Why do the northern peoples disappear?] Укрепление гражданского единства и гармонизация межнациональных

отношений в Санкт-Петербурге [Strengthening civil unity and harmonization of interethnic relations in St. Petersburg]. http://nacionalsoglasie.kmormp.gov.spb.ru/narody-rossii/malye-narody-severa-i-dalnego-vostoka/. Accessed 10 Jan 2017

Batyanova EP, Bakhmazkaya IV, Vlasova IV, Poplavskaya XP (2009) Narody Rossii. Atlas Kultur i Religii [Peoples of Russia. Atlas of cultures and religions]. (In Russian). Feoria, Moscow

ВСЛУХ.RU News website (2005) 5th Congress of indigenous peoples of the North, Siberia and the far East of the Russian Federation. http://www.vsluh.ru/news/economics/45942. Accessed 10 Jan 2017

Bykovskii V (2013) Indigenous peoples and industrial development of yamal. http://professor-bykovsky.ru/?page_id=441. Accessed 20 Dec 2016

Federal Law (1996) Об основах государственного регулирования социально-экономического развития Севера Российской Федерации [On the basis of state regulation of social and economic development of the Russian North] SZRF: 26 (3030)

Federal Law (2001) О территориях традиционного природопользования коренных малочисленных народов Севера, Сибири и Дальнего Востока Российской Федерации [On the territories of traditional resource management of indigenous peoples of the North, Siberia and Far East of Russian Federation]. SZRF: 20 (1972)

Federal Law (2011) О праве граждан Российской Федерации на свободу передвижения, выбор места пребывания и жительства в пределах Российской Федерации [On the freedom of movement, choice of place of residence within the Russian Federation for citizens of the Russian Federation]. SZRF: 50 (7341)

Fløistad B (2010) Comparison of Indigenous peoples rights along the arctic routes, CHNL. http://www.arctis-search.com/Comparison+of++Indigenous+Peoples+Rights+along+the+Arctic+Routes. Accessed 20 Dec 2016

Gasprom official site (2008) http://www.gazprom.ru/press/news/2008/april/article56528. Accessed 20 Dec 2016

Gladun E (2015) The development of constitutional norms on indigenous peoples' rights for traditional use of lands and natural resource. In: Rein AM, Gennadi NC, Peeter J (eds) 20 years of new constitutional reforms in Eastern Europe. Eastern European experience. East-West Studies. Journal of Social Sciences of Tallinn University Law School/Tallinn University, pp 113–121

Gladun E, Chebotarev G (2015) Participation of the Northern indigenous peoples in the management of the Russian arctic territories and its legal protection. NISPAcee J Pub Admin Policy 8(1):111–133

Gumilev LN, Kurkchi AN (1989) Ethno-social problems of indigenous peoples of the Russian North, Siberia and Far East: ethno-social alternatives. New Social Development Technology for North, Siberia and Far East. Part 1. (In Russian). Sverdlovsk

International Work Group for Indigenous Affairs (IWGIA) (2016) Indigenous peoples in Russia. http://www.iwgia.org/regions/arctic/russia. Accessed 20 Dec 2016

Judina L (2010) Воссоздание этногенетических портретов [Recreating ethnogenetic portraits]. Наука в Сибири [Science in Siberia] 30–31: 6

Kryazhkov V (2008) Территории традиционного природопользования как формы реализации коренных малочисленных народов на земли [The territories of traditional resource use as a form of implementation of indigenous peoples' right to land]. Gosudarstvo i Pravo 1:44–51

Kryazhkov V (2013) Development of Russian legislation on Northern indigenous peoples. Arct Rev Law Polit 4(2):140–155

Naikanchina A (2010) Indigenous peoples: development with culture and identity. Articles 3 and 32 of the United Nations Declaration on the Rights of Indigenous Peoples. International Expert Group Meeting, New York

News Agency "Arctic-Info" (2016a) Indigenous peoples. http://www.arctic-info.com/encyclopedia/indigenous-peoples. Accessed 20 Dec 2016

News Agency "Arctic-Info" (2016b) Indigenous peoples. http://www.arctic-info.ru/news/chislennost-korennogo-naseleniya-v-yanao-sostavlyaet-pochti-42-tysyachi-chelovek. Accessed 5 Dec 2016

Nuttall M (2000) Indigenous peoples, self-determination and the Arctic environment. In: The Arctic – environment, people, policy. Harwood Academic Publishers, Australia, pp 377–410

Nuttall M (2002) Protecting the arctic, indigenous peoples and cultural survival. Routledge, London

Park E (2008) Searching for a Voice: the Indigenous People in Polar Regions. Sustain Develop Law Policy 8:30

Resolution of the 5th Congress of Indigenous Peoples of the North, Siberia and the Far East of the Russian Federation (2005) Мир коренных народов Живая Арктика [Indigenous peoples world living arctic] 18:18–25

Smorchkova V (2015) Практика взаимодействия коренного населения и промышленных компаний в Арктике [The practice of interaction of indigenous peoples and industrial companies in the arctic]. State Authorities Serv 4(96):63–68

The Constitution of the Russian Federation (2014) http://www.constitution.ru/en/10003000-01.htm. Accessed 10 November 2016

The Foundations of Russian Federation Policy in the Arctic until 2020 and beyond (2009) http://icr.Arcticportal.org/index.php?option=com_content&view=article&id=1791%253. Accessed 3 Dec 2016

Отчетный доклад Президента Ассоциации коренных малочисленных народов Севера, Сибири и Дальнего Востока С.Н. Харючи [Annual Report of the President of the Russian Association of Indigenous Peoples of the North, Siberia and Far East S.N. Kharjuchi] (2009) Мир коренных народов Живая Арктика" [Indigenous peoples' world living arctic] 22:4–19

Устав (Основной закон) Ханты-Мансийского автономного округа – Югры [Charter of the Khanty-Mansi Autonomous Area - Yugra](1995) No. 4-o:62

Устав (Основной закон) Ямало-Ненецкого автономного округа [Charter of the Yamalo-Nenets Autonomous Area] (1998) No. 56-ZAO:12

Chapter 15
The Arctic Journey: Design Experiments in the North

Satu Miettinen and Titta Jylkäs

Abstract Arctic journey was an experiential exhibition series that was realized as a part of HumanSee research project at the University of Lapland in 2016. The exhibition series sought to experiment the multisensory ways of presenting personal experiences in the Arctic and to invite exhibition visitors into the co-creation of the installations. By exploring the exhibition formats, this article asks what is the role of co-design process in the construction of an arctic experience. Two of the exhibition cases included co-creation sessions, where the creation of the arctic experience was taken into a closer experimentation through the co-design process. Through qualitative content analysis and discourse analysis, the research findings suggest that arctic experience, even when not connected to personal experiences in the actual arctic region, is a reflection of the personal understanding of the marginal context that brings the stories of an individual into the core of the experience.

15.1 Introduction

This chapter is focusing on describing the research outcomes of the Arctic Journey, a mobile and experimental exhibition about personal human experiences of the Arctic. The exhibition experiments different formats for concretizing invisible elements of the arctic experience and pursues to create a tangible experience of our arctic lives (Miettinen 2012). The exhibition was aimed at evoking thoughts about how each of us experiences and copes with our living conditions and qualities, and how we relate to others in this sense. The Arctic Journey exhibitions were a platform for creating the arctic experience.

The works of the exhibition were a collection of artwork from HumanSee -research project team at the University of Lapland and from the participatory workshops organized as a part of the HumanSee -project. The HumanSee research project, conducted in 2015–2016, was focusing on service design research around the human experience as a consumer of tourism and retail services. Exhibitions studied

S. Miettinen (✉) • T. Jylkäs
University of Lapland, Lapland, Finland
e-mail: satu.miettinen@ulapland.fi; titta.jylkas@ulapland.fi

© The Author(s) 2017

K. Latola, H. Savela (eds.), *The Interconnected Arctic — UArctic Congress 2016*,
Springer Polar Sciences, DOI 10.1007/978-3-319-57532-2_15

alternative and unconventional formats for exhibitions and co-design processes related to them. The exhibitions tried to locate spaces and formats other than conventional exhibition arenas such as art galleries. This chapter asks what is the role of co-design process when constructing an arctic experience?

The art works incorporated both new media and more traditional media both in digital and printed form, as well as in paper and textile materials. The exhibitions aimed to gather information from the exhibition visitors on how these medias could be used as service design probes or other participatory means to create an arctic experience. The use of different medias enabled to study the online and person-to-person interaction, as well as the auditory part of the user experience.

The Arctic Journey – Design Experiments in the North exhibition toured in six locations including some of the main art and design universities: Parsons the New School for Design, New York; Finlandia University, Hancock, MI; Emily Carr University of Art + Design, Vancouver British Columbia; Stanford University Campus in Palo Alto and University of California, Berkeley Campus. The exhibitions were important means for visualizing and producing sensory scapes to illustrate an arctic experience.

The Arctic Journey was an artistic research project where the exhibition design and implementation process produced the data for this research paper. The exhibition design process included group discussions that were noted and documented (Nimkulrat 2007). The exhibitions were as well documented and discussed. The obtained research data was then analyzed by using qualitative content analysis (Strauss 1987) through categorizing and coding the themes discussed during the project. These themes are later on discussed in the larger theoretical framework by using discourse analysis (Jørgensen and Phillips 2002). In artistic research process, it is the process itself that helps the researcher to discover the research questions. The artistic production and the overall process stretches out over a longer period, where new discoveries are made and the research phenomena becomes more focused and explicit. Artistic research is many times trying to transform tacit knowledge into explicit and experiential knowledge (Nimkulrat et al. 2016).

15.2   Designing the Arctic Journey

The design process for the exhibition took place through a series of design meetings where the exhibitions were discussed. The exhibition team, consisting of the HumanSee project team members, discussed how to produce the arctic experience that would not only limit to the visual aspects of the artwork, but that would include all the senses and even the cultural context and the behavior experienced in the exhibition. As all the exhibition locations were different, the design team came up with a concept that could be transformed into these different circumstances and would be flexible to use including interactive, tactile and visual elements.

One of the major exhibition elements was an interactive multimedia installation. This included a series of arctic photography that exhibition visitors could use to

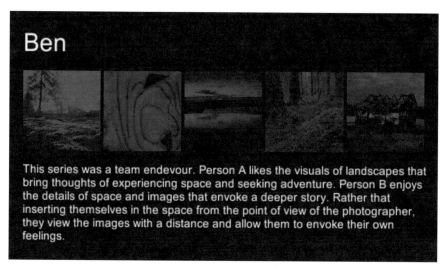

Fig. 15.1 One story in the digital Arctic journey picture collage

create their own arctic experience. The pictures had been chosen by the exhibition team and the selection was guided by their own experiences of the Arctic. The pictures were then re-iterated and formulated by the exhibition visitors into their own stories through a computer interface (Fig. 15.1). All the created visual journeys were then visible for the next visitors, and the collection grew throughout the exhibition series.

The audio experience of the Arctic was presented in the exhibition as garment installation pieces: there were four different types of hats, representing the four different arctic seasons. Each hat was equipped with headphones and audio players. The hats repeated sounds recorded around the Lapland region representing moments from everyday life. The exhibition visitors were encouraged to wear the hats and to imagine what the sounds in the recordings could mean to them.

One of the interactive elements of the exhibition was a paper installation consisting of paper cut snowflakes positioned in the space as a hanging composition. Exhibition visitors were encouraged to create their own snowflakes by following the easy visual instructions and they could place their creation as a part of a co-created installation. The cutting of snowflakes is traditional crafts that many children are involved with, and such a simple form of crafts was familiar for many visitors. This form of installation therefore brought a historical and narrative aspect to the exhibition.

To introduce the tactile level to the exhibition, the exhibition visitors were provided woolen socks to be used during their visit to the exhibition. Woolen socks are an important part of the arctic experience as they represent the practical side of coping with the coldness, but also carry memories of warmth and softness. Knitted woolen socks in varied patterns are also a very traditional form of handicraft in the arctic culture.

The visual elements of the exhibition also included pictures of the arctic region, paired with pictures from similar visual settings in another context. Half of the pictures were photographed at the Arctic Circle around Rovaniemi, Finland, and their counterparts were photographed in New York City, USA. By making the contrasts strongly visible, the picture pairs illustrated the dichotomy of the marginal and the hub. They illustrated the remoteness and isolation of the Arctic against the crowded centers and traffic of a metropole.

The pictures were supported by other visuals such as graphic signs that gave the visitors information and introductions to the exhibition, but also made the familiar street signs appear in a different form when adding an arctic perspective in them. Also in two of the exhibition locations, at Stanford University and University of California Berkeley, posters with QR codes were placed in public places at the university campus areas to quide the viewers to the digital content of the exhibition.

15.3   Co-design in the Exhibition Cases

Two of the exhibitions were combined with collaborative processes through workshops; first one in New York and the second in Hancock, Michigan. The exhibition and workshop in New York were organized at Parsons the New School of Design and the participants were both students and professionals. The exhibition here included the most variety of elements such as an interactive multimedia installation, hats as sound garments, picture pairs, collaborative snowflakes, woolen socks and graphic signs. In addition to this, the exhibition visitors participated in a video production workshop where they created their own arctic journey by using material that they had shot beforehand in New York. Through videos, they were interpreting their view about the arctic experience and created short stories about their journeys.

The second exhibition was conducted at the Finlandia University in Hancock, Michigan. The associated workshop included five parts. In the first part, the workshop participants were asked to complete a storytelling probe where they shared their life histories and experiences with Satu Miettinen and Titta Jylkäs. Collaborative or participatory processes require probing to stimulate interaction between the facilitator and participants.

The core to probing is in creating contextual understanding (Gaver et al. 1999; Mattelmäki 2008). This method of using a probe was used in the 'Wings to Fly' collaborative art process that was initiated by sending storytelling probes to the youth participants. These storytelling probes helped the workshop facilitators to understand and relate to the life stories of the participants as well as to understand the challenges they were facing. The narrative practice is a powerful mean to open communication and make sense of societal contexts and everyday lives (Ryan 2004). The life stories were then analyzed by qualitative content analysis categorizing and coding themes that were afterwards discussed by Satu Miettinen and Titta Jylkäs. The second part of the workshop was a narrative practice where the workshop

Fig. 15.2 Wings to Fly (*left*) and paper mandalas (*right*) as results of co-design workshops at Finlandia University, Hancock (Photograph by Titta Jylkäs)

participants were first writing words around different themes related to their life histories and then sharing a story, narrative, around this.

In the third part of the process, these previously obtained narratives were transferred into a textile installation that was collaboratively created by the workshop participants. The participant stories were visualized and painted on feathers made from paper and textile. The feathers were sewn together to form a wing resulting as "Wings to Fly" installation (Fig. 15.2) in Reflection gallery in Hancock, Michigan, where it was presented for some months.

As it was experienced in the workshop with the youth, artistic production is a way to process significant personal histories, experiences and decisions. Art processes facilitate identity construction, permitting the reconciling of multiple identities, fractured selves and personal stories to guide individuals and groups in coping with life's realities (Miettinen et al. 2016).

The goal of the installation process was to give the participants artistic, visual and verbal tools for processing their life histories and finding empowering elements in their stories. Paulo Freire et al.'s (2000) discussion of education as an intervention and social action where art functions to strengthen self-expression, creative ability and learning experience was one of the motivations of the project. Freire et al. (2000) discusses how art and artistic process gives means to participate, 'to do with', rather than 'for'.

The fourth part of the process was a video workshop where the participants were creating a short video about their process to create "Wings to Fly" together. As the fifth and last part of the exhibition process a collaborative workshop was organized around storytelling at the local indigenous Ojibwe Community College. A storytelling practice started the workshop and it concluded by creating small paper mandalas to visualize the stories. The stories were shared by the authors to the other participants. Some of them were true empowerment stories that described surviving from hard personal challenges like subsistence use. These stories created a mandala installation of the everyday coping stories (Fig. 15.2). The storytelling method was inspired by the imagery weave practice of Chueng-Nainby et al. (2014).

15.4 Findings from the Arctic Experience

In the exhibitions, the arctic experience was created by using both co-creational and visual elements. Parsons the New School of Design offered an arena for the use of different visual, audio and tactile elements. Visual dichotomies, soundscapes and the feel of woolen socks were providing the elements of everyday experience in the Arctic context. These elements were aiming to construct something that the exhibition visitors could recognize and relate to, rather than something exotic and strange. The exhibition was creating a forum where to discuss different everyday elements related to the arctic experience.

Co-design was an important part of the arctic design process. The exhibitions included parts that were collaboratively created either in the exhibition situation or in workshops. Participation and inclusion seemed to be one of the nominal themes of the arctic design.

The series of exhibitions showed that creating and sharing an experience about the Arctic is both personal and collaborative. The experience is subjective, but as it is shared and created with others, it provides a person with the possibility for reflection and realization. The used co-design process supported both individual exploration and collaborative sharing, when the stories could be iterated and shared.

References

Chueng-Nainby P, Fassi D, Xiao DJ (2014) Collective envisioning with local community for village regeneration at inner Mongolia. In: book chapter of emerging practices: design research and education conference
Freire P, Ramos MB, Macedo D (2000) Pedagogy of the oppressed, 30th edn. Continuum International Publishing Group, New York
Gaver B, Dunne T, Pacenti E (1999) Design: cultural probes. Interactions 6(1):21–29
Jørgensen MW, Phillips LJ (2002) Discourse analysis as theory and method. Sage Publication, London
Mattelmäki T (2008) Probing for co-exploring. CoDesign 4(1):65–78
Miettinen S (2012) Service design, radical innovations and arctic wellbeing. In Tahkokallio, P. Ed. Arctic Design. Opening the Discussion. LUP
Miettinen S, Akimenko D, Sarantou M (2016) Narratives creating art for empowerment. 'MEDIATIONS: art & design agency and participation in public space' Royal College of Art, London on the 21st–22nd of November, 2016

Nimkulrat N (2007) The role of documentation in practice-led research. J Res Prac 3(1):6

Nimkulrat N, Niedderer K, Evans M (2016) On understanding expertise, connoisseurship, and experiential knowledge in professional practice. J Res Practice 11(2):1

Ryan ML (ed) (2004) Narrative across media: the languages of storytelling. University of Nebraska Press, Lincoln

Strauss AL (1987) Qualitative analysis for social scientists. Cambridge University Press, Cambridge

Chapter 16
The Bicycle and the Arctic – Resilient and Sustainable Transport in Times of Climate Change

Alexander Meitz and Karoline Ringhofer

Abstract In Alaska, alternative transport modes to cars and other motorized, petrol powered vehicles are gaining in importance due to increasing urbanization and as adaptive responses to risks of infrastructural damage to transportation networks in facing the climate change. Bicycling functions as a transportation practice in response to increasing infrastructure disruptions, offering a strategy towards sustainable and resilient means of transportation in the times of global climate change and its associated challenges. Changing weather conditions and demographic changes lead to the necessity to establish more adapted infrastructures. The use of the bicycle hereby offers a suitable mode of transport in Arctic and Subarctic areas. New technologies such as *fat bikes* make bicycling throughout the winter season possible. The combination of the theoretical concepts of sustainability and resilience bears the potential to foster concrete solutions and action strategies for policy makers with respect to infrastructural challenges in the Arctic.

16.1 Introduction

Contrary to popular opinion, the bicycle is not a new phenomenon in Alaska. During the gold rush at the turn of the last century (1898–1908) it was used as a means of transport not only in commuting trips but also in long distances like Dawson (Yukon, Canada) to Nome (Alaska, USA), covering more than 1200 km/772mi (Cole 1985). More than a 100 years later, in summer season, cyclists are spotted frequently commuting in Alaska's urban areas and using the bicycle as leisure activity. Moreover, even in Alaska's particular environmental conditions, winter cycling is becoming

A. Meitz (✉) • K. Ringhofer
Department of Social and Cultural Anthropology, University of Vienna, Vienna, Austria
e-mail: alexander.meitz@univie.ac.at; karoline.ringhofer@gmail.com

© The Author(s) 2017
K. Latola, H. Savela (eds.), *The Interconnected Arctic — UArctic Congress 2016*,
Springer Polar Sciences, DOI 10.1007/978-3-319-57532-2_16

increasingly popular not only as sports and recreation, but also as commuting device due to newly developed technologies, new bicycle lanes separated from car traffic, and federal grant programs such as the Safe Routes to School (SRTS) or the AMATS[1] Bike Plan Implementation Project (CRW Engineering Group 2017).

This chapter presents preliminary results from an ongoing research that focuses on urban bicycling in Alaska. The research design includes interviews with bicycle commuters, policy makers, bicycle advocacy organizations and many hours of participant observation on wheels, immersing in the field and directly experiencing the particular conditions of cycling in Anchorage and Fairbanks in both summer and winter. Additional research would ideally incorporate interdisciplinary collaboration, as changing conditions in the Arctic and Subarctic require "knowledge of geomorphological processes … and of infrastructure design and function" (Hovelsrud et al. 2010), as well as an anthropological perspective, giving attention to the local population's vulnerability and adaptive capacity towards these changes.

16.2 Theoretical Framework

Bicycling is frequently linked with notions of sustainability due to its energy-efficiency and the fact that its use doesn't produce any carbon emissions. The concept of sustainability itself refers to "the long-term ability to continue to engage in a particular activity, process or use of natural resources" (Benson and Craig 2014, 778). Its main drawback however is that this concept relies on a predictable economic and social development and that it is based on the assumption that there exist maintainable equilibrium states of socioeconomic and ecological systems within the earth's carrying capacity. An example which illustrates this notion stems from sustainable forest management, stating that one should not cut down more trees than can grow back over the same period of time. Yet, in times of climate change our world is confronted with uncertainty. Systems are in process of adaptation and transformation. Global climate change, the loss of biodiversity and the constant increase of the per capita consumption of resources create a high probability of rapid, dynamic and non-linear changes in social and ecological systems (Benson and Craig 2014, 777). Non-linearity refers to the possibility that climate processes can abruptly accelerate or slow down. Due to this uncertainty, it is argued here, that sustainability has to be combined with a resilience-approach in order to have an adequate analytical tool at hand in the face of future challenges. Resilience is a concept to account for the capacity of a system to cope with change, i.e. stress on the respective system. It is a "measure of the amount of change the system can undergo and still retain the same controls on function and structure" (Berkes and Jolly 2002). Adaptive capacity as it is used here "refers to the underlying capacity to adjust to changing conditions, it can be considered an important expression of resilience" (Arctic Resilience Report 2016, 9). The concept of resilience alone however has the disadvantage that also unsustainable system configurations may prove

---

[1] Anchorage Metropolitan Area Transportation Solutions

to be resilient. Thus, it is the combination of both concepts which allows for a long-term sustainable solution which also buffers from external shocks. In the present research, aspects of sustainability and resilience as discussed regarding the use of the bicycle in Alaska's urban areas.

16.3   The Bicycle in Urban Alaska

The relevance of sustainable and resilient modes of transport in Alaska's urban areas stems from the trend towards increasing urbanization in the region. "In urban areas, the demand for pedestrian, bicycle, and transit travel is increasing and this is expected to continue as the State's population becomes more concentrated in urban areas" (Facilities 2016). Geophysical processes of climate change expedite these migratory movements (Hovelsrud et al. 2010). As more and more people are moving to urbanized areas, questions about mobility and infrastructure[2] become prevalent for decision makers (Norden 2011). In 1995, the Department of Transportation and Public Facilities (DOT&PF) presented its first bicycle and pedestrian plan which is an integral part of the 'Vision: 2020 Alaska's Long-Range Statewide Transportation Plan'. Now, 22 years later, an updated version is available. The policy plan 'Let's keep moving 2036' addresses increasing bicycling demands in urban areas.

It is argued here that the bicycle offers an adaptive strategy towards increasing urbanization in Alaska and future climate-induced challenges. Although currently only a marginal percentage of less than 2.5%[3] of Anchorage's commuters are using the bicycle as means of transportation to work, it has the potential to offer an interesting alternative for Arctic and Subarctic cities. Being unfamiliar with the topic, one might reason that Alaska's long winters impede the use of the bicycle. Yet, Minneapolis, whose winter conditions are comparable to those of Alaska's urban regions, is a top-ranked bicycle city with almost 6% of the population commuting (United States Census Bureau 2016). In the European Arctic, the percentage of cyclists even is considerably higher. The city of Oulu in Finland and city of Umeå in Sweden both with an overall bicycle modal share at around 20% (EPOMM 2017) show that bicycle infrastructure combined with the provision of public transportation and community attitudes, public support and pro-bike policies are as important factors influencing the use of the bicycle, as topography or climate (Bergström and Magnusson 2003).

Besides political, legal and infrastructural aspects, new technologies facilitate the use of the bicycle in the Arctic. Terrence Cole comments on winter cycling during the beginning of the 20th century:

---

[2] Infrastructure, as it is used here, refers to transportation networks.

[3] The number stems from the 'Regional Household Travel Survey 2014'. An actual number of cyclists is not yet available. The Anchorage Bike Community just installed their first bike counter in September 2016 and in combination with bike-tracking software such as STRAVA they will be able to get more comprehensive data.

In the low temperatures, bearings would freeze and the tires get stiff. A fall on the ice when the temperature was far below zero could easily shatter a pedal or a handlebar, or a knee or an elbow. For good reason most of the men whose stories are included in this book were thought to be a little mentally deranged (Cole 1985).

Even nowadays, non-cyclists react in a similar way if asked for their opinion on people who commute by bicycle in winter. Nevertheless, today's bicycle technology is way more capable to handle temperatures far below zero Fahrenheit. The introduction of *fat bikes* (Fig. 16.1) to Alaska makes bicycling in the winter season possible, as their design enables these off-road bicycles to ride on unstable ground such as mud and snow and therefore being able to use unpaved mountain bike single-tracks[4] for commuting. In year 2000, *fat bike* tires with a width of at least 3.8 inches/97 mm appeared on the market. Reduced tire pressure brings more of the tire's surface area in contact with the ground, allowing the cyclist to ride on soft terrain, and is thus perfectly suited for winter cycling in the Arctic. It is a low-cost solution for citizens in comparison to the costs of a car, but not barrier-free in terms of physical requirements. Congruent with the literature dealing with bicycling in northern cities such as Rotterdam and Vermont (Böcker and Thorsson 2014; Spencer et al. 2013), precipitation, cold temperatures, wind and limited daylight are among the major factors which deter people from using the bicycle all around the year. Northern weather conditions pose obstacles concerning comfort and safety, with an

Fig. 16.1 Fat bike in Fairbanks, Alaska (Photo: A. Meitz)

---

[4]Singletracks are about the width of the bicycle and therefore not accessible for motorized vehicles.

increased risk of accidents due to icy paths, non-existing or badly maintained bicycle infrastructure and darkness. These factors dissuade many cyclists from commuting in winter season, leaving winter cycling to the most entrenched idealists, who perceive all-year bicycling as lifestyle choice. The specific weather conditions add to an understanding why it is unlikely for bicycling to form a major proportion of transportation in Arctic areas. Another disadvantage in its use is its limited capacity of transporting large goods. Although there exist cargo bicycles, winter conditions impede their usage. Cold temperatures, wind and precipitation further pose challenges to carry children in seats and carriages, making bicycling in the Arctic's winter conditions a family-unfriendly undertaking. Even if the bicycle presents an inexpensive mode of transport – especially in comparison to motorized vehicles – the costs to set up the winter equipment (including *fat bike* and outfit) vary between US-$3000–4000, which equals roughly the average monthly income in Alaska.

It is widely known that bicycling presents an energy-efficient mode of transport in the sense that it is human-powered, offering a beneficial cardiovascular exercise (Pucher and Buehler 2008, 496). Its use is not dependent on fuel, bicycle lanes need less space than paved roads for cars, their maintenance costs less money, and needs less resources. Especially for urban areas with increasing traffic, bicycling has the key benefit of producing zero emissions, contributing to improved air quality. Further, it has the potential to reduce noise pollution and promotes an active lifestyle, which reflects itself in the use of the bicycle for recreation and physical activity. Sport lovers seek out unpaved terrain in woodlands and snow conditions, and the tourism sector heavily relies on this perpetuated image of wilderness and adventure.

Analyzing the bicycle's adaptive capacity towards changing conditions provides us with information about this mode of transportation's resilience towards climate perturbations. The root of the resilience of bicycling stems from the fact that cyclists can create and shape their own tracks. They are not as dependent on paved roads as cars, but can move on gravel roads in summer and packed snow trails in winter season, which get created gradually by the cyclists themselves due to the applicability of *fat bikes* on unstable and soft grounds. An interesting aspect here comes with the fact that the creation of packed snow trails is flexible and shows adaptive potential. In case of bad road maintenance like uncleared bike lanes or piles of snow created by snow ploughs, snow conditions enable cyclists (especially those using *fat bikes*) to create their own tracks, shaping them to their necessities. This aspect makes them much less dependent on infrastructure provided by the municipality.

Further, bicycling offers an adaptive response towards widely reported changing weather conditions in the Arctic. In the context of Alaska, these changes manifest themselves in unusually warm temperatures and decreased precipitation (Wendler and Shulski 2009, 300), accompanied by non-linear events such as sudden disruptions like landslides stemming from thawing permafrost grounds, snow avalanches and variability in snowfall, which affect mobility. Snowfall is subject to increased uncertainty because of changing frequency, timing and intensity such as occasional unusually heavy snowfall and wider fluctuations with more frequent alternations of thaw and freeze periods. A longer transition between the seasons with repeated freeze-thaw-cycles makes road maintenance with snow and ice control more complicated and costly. The use of bicycles offers a resilient commuting strategy, as *fat bikes* do not depend on road clearance and are especially designed for snow. This characteristic makes winter cycling an adaptive strategy towards sudden snowfalls, in situations where cars are much more dependent on external help. Further, in thawing cycles *fat bikes* can easily maneuver over muddy terrains. This aspect enables the use of the bicycle in the context of increasing temperatures, as unpaved lanes (Fig. 16.2) remain accessible for cyclists, even in the case of melting snow and soft, wet, unstable soil such as thawing permafrost grounds. For icy roads there exist studded tires, making all-year round cycling possible. Another feature regarding the adaptive capacity of cyclists is their resilience towards road obstructions. If roads get blocked due to unexpected geomorphological events such as landslides or fallen trees, cyclists can easily make their way around obstacles. All these aspects prove the high adaptability of the bicycle in the face of increasing climate variability. All around the circumpolar north, thawing permafrost will lead to increased costs in road maintenance (U.S. Arctic Research Commission Permafrost Task Force 2003). In Alaska "uneven sinking of the ground in response to permafrost thaw is estimated to add between $3.6 and $6.1 billion (10–20%) to current costs of maintaining public infrastructure such as buildings, pipelines, roads, and airports over the next 20 years." (Chapin et al. 2014, 520). Even if roads fall into disrepair, people with *fat bikes* will continue to be able to transport themselves.

Fig. 16.2 An unpaved path with packed snow created by cyclists and pedestrians in Fairbanks, Alaska (Photo: A. Meitz)

Although Alaska's urban regions are not directly affected[5], this aspect is insofar of relevance as melting permafrost is one of the reasons for increasing migration from high-risk areas.

16.4 Conclusion

The trend towards urbanization in Alaska raises questions about mobility, means of transportation and infrastructure. The bicycle offers a sustainable commuting device, reduces traffic, emissions, noise pollution and contributes to better air quality in cities. Its usage is however to be seen with restrictions, as it is neither an option for people with limitations, nor for transporting big masses. Cold temperatures, precipitation and limited daylight hours further deter people from the use of the bicycle due to discomfort and safety concerns. Because of these disadvantages, it might never form a major proportion of transportation in Arctic winter conditions, but offers certain adaptability advantages towards changing weather conditions in comparison to car users. This aspect constitutes the bicycle's resilience factor, as its use is less dependent on paved and well-maintained roads and it can ride on unstable terrain such as snow and mud. Bicycles have more flexible maneuverability than cars – and that quality can be particularly useful in the Arctic with its more variable and changing weather conditions. Due to its applicability on various terrains, the use of the bicycle makes it possible to circumvent road obstructions, and to be less dependent on road maintenance. The local practice of all-year cycling in other Arctic and Subarctic regions with comparably high levels of cycling due to well developed bicycle infrastructure and pro-bike policies can be instructive for policy makers in Anchorage and Fairbanks. Moreover, bicycling in the north presents an attractive leisure activity and forms part of the eco-tourism in the Arctic. In the long run, bicycle infrastructure and the idea of a complete street design can help city governments to create an infrastructure with safe access for all users. Bicycling as a sustainable and resilient means of transportation is – against expectations – a feasible option for commuters in the circumpolar north.

References

Benson MH, Craig RK (2014) The end of sustainability. Soc Nat Resour 27:777–782. doi:10.108 0/08941920.2014.901467
Bergström A, Magnusson R (2003) Potential of transferring car trips to bicycle during winter. Transp Res A Policy Pract 37:649–666. doi:10.1016/S0965-8564(03)00012-0
Berkes F, Jolly D (2002) Adapting to climate change: social-ecological resilience in a Canadian western Arctic community. Conserv Ecol 5:18

---

[5] Fairbanks area is not a high hazard zone, with only discontinuous permafrost, but future disruptions may occur such as road damages due to moving grounds.

Böcker L, Thorsson S (2014) Integrated weather effects on cycling shares, frequencies, and durations in Rotterdam, the Netherlands. Weather Clim Soc 6:468–481. doi:10.1175/WCAS-D-13-00066.1

Chapin FS III, Trainor SF, Cochran P, Huntington H, Markon C, McCammon M, McGuire AD, Serreze M (2014) Ch. 22: Alaska. Climate Change Impacts in the United States: The Third National Climate Assessment, In: Melillo JM, Richmond T, Yohe GW (eds), U.S. Global Change Research Program, pp. 514–536. doi:10.7930/J00Z7150

Cole T (1985) Wheels on Ice – Bicycling in Alaska 1898–1908. Alaska Northwest Publishing Company.

CRW Engineering Group L (2017) AMATS. http://www.anchoragebikeplan.com. Accessed 02.01.2017

EPOMM (2017) TEMS - The EPOMM Modal Splitt Tool. http://www.epomm.eu/tems/cities.phtml. Accessed 01.02.2017 2017

Facilities TaP (2016) Let's keep moving 2036: policy plan. Alaska Statewide Longe-Range Transportation Plan

Hovelsrud GK, White JL, Andrachuk M, Smit B (2010) Community adaptation and vulnerability integrated. In: Hovelsrud GK, Smit B (eds) Community Adaptation and Vulnerability in Arctic Regions. Springer, Heidelberg, pp 335–348

Norden (2011) Megatrends. Team Nord, 2011 edn. Danish Ministry of Foreign Affairs, Copenhagen. doi:http://dx.doi.org/10.6027/tn2011-527

Pucher J, Buehler R (2008) Making cycling irresistible: lessons from the Netherlands, Denmark and Germany. Transp Rev 28:495–528. doi:10.1080/01441640701806612

Spencer P, Watts R, Vivanco L, Flynn B (2013) The effect of environmental factors on bicycle commuters in Vermont: influences of a northern climate. J Transp Geogr 31:11–17. doi:10.1016/j.jtrangeo.2013.05.003

U.S. Arctic Research Commission Permafrost Task Force (2003) Climate change, permafrost, and impacts on civil infrastructure. Special Report 01-03, U.S. Arctic Research Commission, Arlington, Virginia

Wendler G, ShulSki M (2009) A century of climate change for Fairbanks, Alaska. Arctic:295–300

Part III
Building the Long-Term Human Capacity

Chapter 17
Human Capital Development in the Russian Arctic

Alexandra Kekkonen, Svetlana Shabaeva, and Valery Gurtov

Abstract The article presents results of human capital status in the Russian Arctic and proposed measures to improve the situation. The main method for improvement is seen as the popularization of in-demand occupation list through federal web-portal to increase the balance between education system output and the labour market needs. Development of the Russian Arctic human capital is a strategic priority for increasing economy's competitiveness in terms of globalization as well as considering raising interest and attention to this topic. There is an objective contradiction in the Russian Arctic development: it is expensive to explore and develop the Arctic, but the Arctic is a territory full of possibilities and resources. Today's priority is the economic development. This will lead to development of labour market and to social aspects progress. As a result, the severe life conditions in the Arctic and health deterioration of population could be improved.

17.1 Introduction

Development of the Russian Arctic human capital is a strategic priority for increasing economy's competitiveness in terms of globalization as well as considering raising interest and attention to this topic. The role of the regions' human capital development also increases the need to pay attention to their territorial characteristics and special features of economic, natural, geographical, social-demographic and other factors, based on regional strategies devoted for self-development (The development strategy of the Russian Arctic and national security for the period until 2020 2008).

The condition of human capital in the Russian Arctic is an important part of the territory's strategic development. The modern society – irrespective of the territorial position – is facing number of challenges like globalization, population ageing, and decrease of labor productivity level. Problems in the society caused these challenges lead to deterioration of the population's social-economic status, marginalization of

A. Kekkonen (✉) • S. Shabaeva • V. Gurtov
Petrozavodsk State University, Petrozavodsk, Russia
e-mail: alexandra.kekkonen@gmail.com

© The Author(s) 2017
K. Latola, H. Savela (eds.), *The Interconnected Arctic — UArctic Congress 2016*,
Springer Polar Sciences, DOI 10.1007/978-3-319-57532-2_17

youth, and imbalance of occupational and qualification structure. The described problems should be overcome and it is important to evaluate them and monitor their dynamics. The fore mentioned issues strongly affect the human capital, because life in Northern territories has great impact on health (one of important components of human capital), while for other territories this factor is not crucial (Agenda 21 1992).

Long-term human capacity building in Russian Arctic includes mainly human resource development measures, concentrating on labour market and education system balance.

17.2   Russian Arctic: Long-Term Human Capacity Building

At present, the Russian Arctic (Russian Federation Presidential Decree 2014), occupies a central place in its importance, serving strategic national interests in economics, geopolitics and resource base. The Russian Arctic includes the following regions: Arkhangelsk region, Vologda region, Murmansk region, the Republic of Karelia, Komi Republic, Nenets Autonomous region, Yamalo-Nenets Autonomous region, Krasnoyarsk region, Chukotka Autonomous region, The Republic of Sakha (Yakutia). The share of the gross regional product produced in the Russia Arctic zone, in the total gross regional product of the Russia regions in 2014 was 5.2% (The official statistic data about social and economic development Russia Arctic Zone 2017). Arctic territory of the Russian Federation is a large-scale economic zone, including some of the largest economic centers in the field of mining, including fuel and energy, shipbuilding and navigation, as well as fishing.

Arctic is a strategically important – yet challenging – region. Particular attention is being given to building long-term human capacity in the region, but with weak and imbalanced economy it is difficult to carry out complex development. Capacity building is much more than education and training and includes (Agenda 21 1992) human resource development, the process of equipping individuals with the understanding, skills and access to information, knowledge and training that enables them to perform effectively. It also includes organizational development, the elaboration of management structures, processes and procedures, not only within organizations but also the management of relationships between the different organizations and sectors (public, private and community). Another dimension is institutional and legal framework development, making legal and regulatory changes to enable organizations, institutions and agencies at all levels and in all sectors to enhance their capacities. Russian Government at present concentrates mainly on skills development and access to information.

Economists in Russia treat human capital as scope of knowledge, skills, abilities, motivations, which are used to meet the diverse needs of individuals and society as a whole. This scope of characteristics are related to obtaining income resulting from investments in the development of professional skills of its holder, as well as its efficient use during consumption in the human capital reproduction – recurrent processes of re-creation. The efficiency of human capital is the impact and the profit which the individual brings to society as well as the impact on personal development, and is caused by the relevant occupation and qualification structure (Kekkonen 2013). Consequently, human capital dynamics is considered through revealing demographic structure with the emphasis on labour force status.

The Russian Arctic is home to 2.4 million people that are 1.6% of the total Russian population (The official statistic data about social and economic development Russia Arctic Zone 2017). Table 17.1 presents demographic aspects of the human capital development in the Russian Arctic. The data shows that almost all indices are lower in comparison to Russia as a whole. This means that there is a serious threat to human capital reproduction in Arctic region, which should be kept in mind.

The Arctic is dominated by younger population compared with Russia (Table 17.2). Elderly people leave the Russian Arctic area after retirement, and move to a place with more favorable climatic conditions[1].

Population structure is characterized by a high proportion of the economically active population – average value in areas of the Arctic region is 71.8% against 69% for Russia as a whole. The employment rate in the Russian Arctic also exceeds the average and amounted to 69.5% in 2014, while all-Russian level counting 65.4% (The official statistic data about social and economic development Russia Arctic Zone 2017). Situation on the labour market in general can be characterized as possibility and strength for the future human capital development.

Analysis of existing interpretations of the 'human capital' notion has shown that, historically, education is the main and decisive element in the structure of human capital. Education is the basis for the formation of knowledge and skills – human competencies – for the further implementation of the labor market. Elements such as quality of life, culture and health are considered by researchers as auxiliary elements in the formation of human capital (Kekkonen 2013). Life in Northern territories has great impact on health, while for other territories this factor is not crucial (Belisheva and Petrov 2013).

Structure of the Human Development Index and its main components[2] in the regions of the Russian Arctic shows that the region is below the all-Russian level. The exception is the Education Index which is higher in most regions nationwide. Also the Income Index in rich regions such as the Republic of Sakha (Yakutia) and Chukotka Autonomous region, Komi Republic is significantly higher than the all-Russian level. This is due to the presence of rich natural resources and their exploitation.

Specific features related to life in the Arctic (Belisheva and Petrov 2013) can cause an increase in the incidence of circulatory system, tumors, poisoning, and neurological disorders. In addition to this, statistics show that living in the North increases the number of injuries and accidents at work (The official statistic data

---

[1] Retirement age in Russia is 60 years for men and 55 years for women. In Russian Arctic figures are 55 and 50 years respectively.

[2] https://en.wikipedia.org/wiki/Human_Development_Index

Table 17.1 Demographic structure of the Russian Arctic population

	Total fertility rate – The number of born per 1000 population	Life expectancy at birth in 2014, total	Life expectancy at birth in 2014, men	Life expectancy at birth in 2014, women	Infant mortality rate – The number of children dying before the age of 1 year per 1000 live births (2014)	The immigration rate per 10,000 population (2014)
Arkhangelsk region	12.5	70.2	64.16	76.32	6.8	−68
Vologda region	13.6	69.74	63.66	75.93	7.6	−7
Murmansk region	11.7	69.97	64.02	75.72	6.4	−65
Republic of Karelia	12.3	69.36	62.99	75.69	6.7	−7
Komi Republic	14.2	69.05	63.05	75.12	5.1	−107
Nenets Autonomous region	16.8	70.65	64.72	76.21	5.5	1
Yamalo-Nenets Autonomous region	16.9	71.92	67.02	76.86	8.4	−112
Krasnoyarsk region	14.4	69.23	63.6	74.83	8.3	3
Chukotka Autonomous region	13.7	62.32	58.84	66.62	23.4	−30
The Republic of Sakha (Yakutia)	17.8	69.81	64.34	75.5	8.0	−70
Russia	13.3	70.93	65.29	76.47	7.4	19

The official statistic data about social and economic development Russia Arctic Zone (2017)

Table 17.2 Labour force status in the Russian Arctic and Russia on the whole

	Russia Arctic zone	Russia
Distribution of population according to age (Jan.1 2014)		
Below working age	19%	17%
Above working age	17%	24%
In working age	64%	59%

The official statistic data about social and economic development Russia Arctic Zone (2017)

about social and economic development Russia Arctic Zone 2017). As a result, statistics shows that the incidence of a disease rises in all Arctic regions.

Human capacity building is a complex process that should take into account the strengths and weaknesses of the state of the regional human capital, and therefore it is necessary to develop a comprehensive policy in this field. There are some areas that will have a multiplier effect on all the human capital of the region, if properly managed. Thus, the development of financial well-being – namely, success in the labor market – will lead to the improvement of human life in general, and improving the state of the economy in particular. Thus, there is more opportunity to neutralize the effects on the health of both the individual (through better medical and prevention measures) and through the allocation of more funds to national and regional programs to improve the lives and health of the population.

17.3   Human Capacity Building Through Labour Market Development

Human capacity building, economic development and the labor market are closely linked. In this area the Russian Arctic has its own specific features.

Despite the high level of economic activity and employment, there are two serious threats in the development of human resource potential of the territories of the Russian Arctic. These are high-migration and the imbalance of supply and demand of labor in regional and professional sections (both in quantitative and qualitative parameters) (Shabaeva et al. 2016). In this manner, "education, providing of training, retraining and advanced training of specialists for work in arctic conditions, taking into account current and projected needs marine geology, hydrocarbon production and processing, marine biotechnology, ICT and other disciplines" are crucial measures to implement (The development strategy of the Russian Arctic and national security for the period until 2020 2008).

On one hand, the Russian Arctic, despite the harsh climate, has a sufficient labor potential and its performance on specific indicators exceed national average values. On the other hand, the labor market status of the Russian Arctic shows that there are problems of supply and demand balance, i.e. there is a significant reserve for a more efficient use of labor resources. Main characteristics of social tension in the labor market are manifested in the shortage of labor, and at the same time as a high proportion of unemployed people with vocational education and young age (Korchak 2015). The revealing of in-demand jobs for Arctic development priorities is one possible solution to find balance.

Economic development and implementation of strategic projects are impossible without appropriate employee supply. Identifying occupations in demand in the region is the foundation for the implementation of the regional policy of population employment and educational policies in the territories of the Russian Arctic.

Economic specifics of the Russian Arctic stipulates the requirements for the education system and authorities in the field of labor and employment to fill staff shortages and cover staffing needs of the territory.

On the basis of a comparative analysis and the structure of training specialties, vocational training programs, occupations according to the statistics of the Russian Arctic territories, a basic list of in demand occupations was developed (Shabaeva et al. 2016). Designated list of occupations for the Arctic is the basis for adjusting the training structure, the opening of new specialties and areas of training, educational resources in other Russian regions and interregional attraction and foreign migrant workers.

For the implementation of many important development projects in Arctic (The development strategy of the Russian Arctic and national security for the period until 2020 2008), timely and accessible information to support projects is highly important. It is necessary to establish public awareness about the in-demand occupations. For this aim, federal web-portal "Staffing Needs for Russian Arctic Zone Development"[3] was developed and is supported (until now) on the behalf of Ministry of Education and Science of Russian Federation.

The main tool for provision of supply and demand balance in the labour market is the list of most popular occupations of Russian Arctic and their detailed description. This list is available on the web-portal. This is concretizing the proposed and approved solutions in the management of Russian Arctic human capital development (Shabaeva et al. 2016).

17.4   Conclusions

Human capacity building includes multiple aspects, and development of human capital is one of them. The Russian Arctic is a complex area for research, development consideration and management. There are both strong and weak sides of the Russian Arctic human capital status. On one hand, there are demographic threats, connected mainly with life conditions and characterized by low development indexes. On the other hand, the economic development of the human capital is a strategic priority for increasing the economy's competitiveness in terms of globalization and at the same time the region has great potential for that. Balanced labour market and economic development will lead to development of labour market and in turn, to progress in social aspects. As a result, severe life conditions in the Arctic and the health deterioration of its population could be improved.

Acknowledgments The research was prepared with support from the state task of Russian Ministry of Education and Science No.30.207.2016/HM, Russian Humanitarian Science Foundation project № 15-02-00231 and The Russian Foundation for Basic Research project № 16-46-100923\16.

---

[3] http://arctic.labourmarket.ru

References

Agenda 21 (1992) United Nations conference on environment & development. Rio de Janerio, Brazil, June 1992. 531 pp

Belisheva NK, Petrov VN (2013) The Murmansk region's population health when implementing the strategy of the development of the Russian Federation's Arctic zone. Proc Kola Science Centre 6(19)

Kekkonen AL (2013) Human capital of the region: the formation and development in modern conditions. Dissertation. Institute for Regional Economics of the Russian Academy of Sciences, St. Petersburg

Korchak EA (2015) Living standards of the North and Russia Arctic regions population. Basic Research 7(3):605–609

Russian Federation Presidential Decree (2014) «On the Russian Arctic Zone land areas» Available via DIALOG: http://base.consultant.ru/cons/cgi/online.cgi?req=doc;base=LAW;n=162553. Accessed 9 Jan 2017

Shabaeva SV, Fedorova EA, Stepus IS (2016) Occupations in demand in Russia Arctic zone as the reflection of macroregion economy development priorities. Econ Manag Probl Solut 7(1):104–117

The development strategy of the Russian Arctic and national security for the period until 2020 (2008) Available via DIALOG: http://government.ru/media/files/2RpSA3sctElhAGn4RN9dHr tzk0A3wZm8.pdf. Accessed 9 Jan 2017

The official statistic data about social and economic development Russia Arctic Zone. Available via DIALOG: http://www.gks.ru. Accessed 9 Jan 2017

Chapter 18
Impact of Wages on Employment and Migration in the High North of Russia

Marina Giltman

Abstract The research presented here examines the impact of wages on employment and migration in the High North regions of Russia. The unique features of the labour markets of these territories include compensative differentials and specific labour protection legislation. It makes the labour supply in the High North of Russia more flexible and labour demand more constrained compared with the rest of the country. Using the regional panel data provided by the Federal State Statistics Service (Rosstat) for the High North regions of Russia from 2005 to 2013, fixed effects models for dependent variables such as the number of employees, number of unemployed and net migration was used to analyse the impact of wages on migration. The obtained estimates demonstrate that wage significantly and positively affects interregional migration to the northern regions. Growth of wages attracts immigrants from other regions of the country and eventually leads to lower wages and higher unemployment in the northern territories. The main findings of this research can be used in the implementation of social policy in the High North regions of Russia.

18.1 Introduction

The High North of Russia (Fig. 18.1) is traditionally associated with an unfavourable climate, remoteness from the European part of the country and high wages, which should compensate for the uncomfortable living conditions. The High North can be characterized by a 5.5% share of the population of Russia and about 10% of the Gross Domestic Product. Most areas of the High North have a strictly oriented economic specialization: in value they produce more than 50% of the product of the extractive industry of Russia. In recent years, despite the multiple growth of real wages in the High North, the number of employees in the majority of cases has demonstrated a negative trend. Thus, according to the Federal State Statistics

M. Giltman (✉)
Tyumen State University, Tyumen, Russia
e-mail: giltman@rambler.ru

K. Latola, H. Savela (eds.), *The Interconnected Arctic — UArctic Congress 2016*,
Springer Polar Sciences, DOI 10.1007/978-3-319-57532-2_18

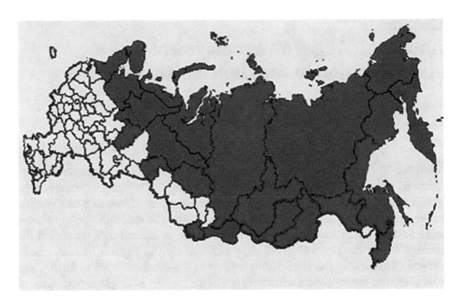

Fig. 18.1 High North Regions of Russia (Source: http://www.gks.ru/bgd/regl/b08_22/IssWWW. exe/Stg/kart.htm with corrections made by the author as of January the 1st, 2014)

Service (Rosstat), wages in 2000–2013 in the High North were 1.6–2 times higher than the average national wage, but the resident population in the High North and equivalent areas in 1999–2014 decreased by 1.207 million people, i.e. by almost 11%. Bignebat (2006), Mkrtchyan and Karachurina (2015), and Sardadvar and Vakulenko (2016) associate out-migration from the High North regions of Russia with labour demand insufficiency.

Demand for labour in the High North regions of Russia is more constrained compared with the rest of the country because of institutional, geographical and economic reasons (Giltman 2016). First of all, the High North region of Russia is unique because of the strongest employment protection legislation among all the regions of the country. Employment protection legislation in the High North regions of Russia is based on Chapter 50 of the Labour Code of the Russian Federation and also implies special consequences for retirement. The Labour Code of the Russian Federation provides benefits for employees in the High North regions such as a regional coefficient of individual wage from 1.15 to 2 times in addition to annual paid leave. Additional employment protection legislation increases the costs to the employer in official hiring, wages and lay-offs, and weakens the enforcement in the High North regions of Russia (Giltman 2016).

Another reason for labour demand insufficiency in the High North regions of Russia is that establishing new industrial production in the High North is extremely costly compared to any other country because of transportation, climate and poor infrastructure (Kryukov and Moe 2013; Pilyasov 2016). Finally, an additional explanation to the demand insufficiency could stem from the one-company towns being the prevalent form of settlements in the High North of Russia (World Bank 2010; Commander et al. 2011; Pilyasov 2016). One-company towns, for example Kirovsk, Onega and Pudozh, are very close to creating a monopsony in the labour market, which constrains labour demand significantly compared with a competitive labour market.

Labour supply also seems to have some special features in the High North regions of Russia. According to the general model of the local labour markets' equilibrium by Rosen (1979) and Roback (1982), local labour markets within one country are connected by migration. An employee selects an area for employment based on nominal and real wages, the productivity of a local economy, local amenities, housing costs in the location and idiosyncratic preferences for the location (Moretti 2011). According to some empirical research (World Bank 2010; Kryukov and Moe 2013; Pilyasov 2016; Giltman 2016; Nalimov and Rudenko 2015; Nazarova 2016; Saxinger 2016; Saxinger et al. 2016), the High North regions of Russia represent themselves as the 'amenity-poor' regions (the term by Greenwood et al. 1991). This means that wages should compensate for poor living conditions and an extreme climate. Thus, in the High North regions of Russia wages should be higher for a worker with the same productivity than in the rest of the country (Coelho and Ghali 1971; Greenwood et al. 1991; Bignebat 2003; Oshchepkov 2015). Empirical evidence for the existence of compensative differentials in the regions of the Russian Federation can be found in the works of Berger et al. 2003; Bignebat 2003; Oshchepkov 2015. Following from this, the migration of employees from the High North to other regions of the country is usually based on the size of wages and not limited by regional amenities due to their absence. The hypothesis that higher wages increase the number of employees both by reducing unemployment and increasing net migration from other regions was tested in this study.

18.2 Results and Discussion

The estimations were based on the regional data provided by the Rosstat from the surveys "Russia's Regions. Socio-economic indicators" and "Economic and social indicators in the High North and equivalent areas" from 2005 to 2013. The observations were taken only from those regions whose territories are entirely included into the High North: the Republic of Karelia, the Komi Republic, the Republic of Sakha (Yakutia), the Tuva Republic, Kamchatka Krai, Arkhangelsk Oblast, Magadan Oblast, Murmansk Oblast, Sakhalin Oblast, Yamalo-Nenets Autonomous Okrug, Khanty-Mansi Autonomous Okrug-Yugra, and Chukotka Autonomous Okrug (see Fig. 18.1).

The regression analyses for migration were conducted first, following the methodology of Vakulenko (2016) by applying a dynamic panel data fixed effects model with spatial effects to estimate net migration among Russian regions. Given that migration is connected with employment and unemployment, they were also estimated in the regressions with the same specifications as the independent variables. It was assumed that changes in wages also affect the number of employees and migrants in the High North with some time lag, because workers need time to internalize the reduction in real wages, to take a decision about leaving and migrating to another region, to organize the move, etc. Employers also do not respond immediately to a change of requirement in the number of employees. They need time to understand the dynamics of wages and labour force population in the region. Three equations for the number of employees, number of unemployed and net migration with the following specifications (18.1) were estimated:

$$
\begin{aligned}
Y_{it} = \ & \beta_0 + \beta_1 lnW_{it} + \beta_2 lnW_{it-1} + \beta_3 lnW_{it-2} \\
& + \beta_4 lnW_{it-3} + \beta_5 lnW_{it-4} + \beta_6 lnW_{it-5} \\
& + \beta_7 lnW_{it-6} + \beta_8 Trade_{it} + \beta_9 Const_{it} \\
& + \beta_{10} Manuf_{it} + \beta_{11} Age_{it} + \beta_{12} lnWomen_{it} \\
& + \beta_{13} GRP_{it} + \beta_{14} Life_{it} + \beta_{15} Search_{it} + \varepsilon_{it}
\end{aligned}
\tag{18.1}
$$

where it represents region i in time t; Y_{it} – dependent variables. In this case the employment, migration and unemployment are connected, because there are almost no any other reasons for individuals to go to the High North regions, except for work and wages. This means that immigrants and native inhabitants can be employed or unemployed. If individuals are unemployed in the long term or they are not satisfied with real wages, they prefer to leave the High North territories. Three dependent variables were estimated to follow this logic: *Emp* – number of employees (1000 people); *Unemp* – number of unemployed (1000 people); *Migr* – net migration (1000 people).

The independent variables were as follows: *W* – average wage per month in a particular region (rubles); *Trade, Const, Manuf* – share of employees in trade, construction and manufacturing with respect to all the employees in the region (%), as a proxy for the structure of the regional economy. The joint share of employment in trade, construction and manufacturing industries, with respect to all the employees in the High North regions, was between 13% and 38%. These industries are present in all the regions, and the share of employees involved in these industries is not correlated with wages and Gross Regional Product (GRP), but correlated with the total number of employees in the region. This is not the case for the extractive industries, which usually have a particular geographic (regional) connection. Age–share of population of working age with respect to the total number of employees in the region (%); *Women* – number of women with respect to 1000 men (persons). *Age* and gender were added because there are special conditions of employment and payment for young people and women working in the High North, and Russian legislation provides for a 5-years earlier retirement for employees in the North. As a rule, not employed pensioners leave the northern territory because of the absence of regional amenities in respect of residence; i.e. the age structure of the labour force population consists almost completely of individuals of working age. *GRP* – Gross Regional Product (1000 rubles), as a variable that describes development of the regional economy (productivity in the model of local labour markets' equilibrium); *Life* – life expectancy at birth (years), as a proxy for regional amenities; *Search* – as the average time taken for job search by the unemployed (months), as a proxy for the search intensity of unemployed job seekers in the regional labour market; ε - normally distributed error term. GRP and wages were deflated by the consumer price index (CPI) for the particular region in the respective years. The number of lags was determined on the basis of the formal Akaike and Bayesian information criterion (AIC and BIC criteria) and tested for the normal distribution of the error term. Estimation was carried out in the Gretl econometric package; the results of the estimation are shown in Table 18.1.

The results showed that wage level is a significant factor that positively affects the involvement of migrants in the workforce in the northern regions. Employment and unemployment react to changes in wages with different lag time and, in both cases, negatively. The results indicate that an increase in wage level attracts new employees from other regions, and this growth of the labour force population leads to the wage declining. More concretely, the estimated coefficients reflect the following picture. A growth in wages of 1% reduces the number of unemployed in the same year to 0.89 thousand people and increases net migration in the same year to 451 people. The simultaneous growth of wages and reduction in unemployment in the region most likely reflects increasing labour demand. In 3 years, the growth in wages by 1% has a negative impact on the number of employees of 0.17%, which reflects the narrowing of labour demand. Also, this ratio can be interpreted as a reduction of real wages after hiring additional employees with a lag of 3 years. Both interpretations demonstrate a reduction in the need of the employer to hire new employees.

At the same time, 3 years after the 1% wage growth, net migration increases by 911 people. The lag looks quite understandable due to the period necessary for information dissemination and making decisions about moving to another region. But, as is shown by the reaction of employment, employers do not have a need for additional workers anymore. In this context, the total decline of unemployment over 5 years by 0.89 thousand people compared with 0.35 thousand people, which has been estimated for the first year, looks quite logical. This difference demonstrates the positive dynamics of the number of unemployed from the second to the fifth years, which affects the signs of the coefficients of the lagged wages, although they are statistically insignificant. Thus, the results of this study reflect a surplus of labour supply with respect to labour demand in the High North of Russia. The growth of the individual labour supply in the High North is too high and leads to a decrease in wages and an increase in unemployment in the northern regions. In contrast, the reaction of labour demand is moderate or at least not so flexible. The lack of flexibility of Russian employers (companies) in the hiring and firing process was also empirically proven by Gimpelson et al. (2010).

Table 18.1 Estimated results of the regression analysis.

Regressor	Regress and coefficient (standard error)		
	lnEmp	Unemp	Migr
Const	−3.13 (5.96)	−1768.8 (1480.8)	−831,308 (994302)
Trade	−0.001 (0.006)	0.78 (2.29)	2336.7 (1141.8)*
Const	0.006 (0.006)	0.85 (1.64)	479.2 (966.9)
Manuf	0.015 (0.015)	2.45 (2.17)	637.9 (2503.9)
Age	0.014 (0.01)	5.03 (3.03)	3764.7 (1846.0)*
Women (log)	1.23 (0.62)*	131.9 (151.0)	9235.8 (100281)
GRP (log)	0.08 (0.15)	0.2 (19.3)	−8632.6 (10685)
W (log)t	0.03 (0.15)	−89.3 (38.4)**	45128.4 (19384.6)**
W (log)t-1	−0.03 (0.1)	47.8 (37.3)	14982.9 (18792.1)
W (log)t-2	−0.1 (0.1)	−1.6 (29.6)	−33464.2 (18725.6)
W (log)t-3	−0.17 (0.07)**	31.1 (28.4)	46048.3 (13454.7)***
W (log)t-4	−0.06 (0.07)	16.2 (25.2)	−7149.5 (12950.1)
W (log)t-5	0.04 (0.14)	54.8 (22.2)**	28093.9 (25047.9)
W (log)t-6	−0.002 (0.1)	−5.1 (30.0)	−34611.5 (19017.6)
Life	0.01 (0.005)**	−0.015 (2.04)	11.8 (843.8)
Search	−0.0007 (0.003)	0.7 (0.9)	123.9 (508.9)
	Number of observations 108	Number of observations 108	Number of observations 108
	Standard regression error 0.013	Standard regression error 3.13	Standard regression error 2077.62
	R^2(within) 0.732	R^2(within) 0.605	R^2(within) 0.809
Joint test on named regressors	$F(15.9) = 1.64057$ p-value = $P(F(15.9) > 1.64057)$ = 0.228844	$F(15.9) = 0.919185$ p-value = $P(F(15.9) > 0.919185)$ = 0.574976	$F(15.9) = 2.53828$ p-value = $P(F(15.9) > 2.53828)$ = 0.0805631
Hausman test	$F(11.9) = 1019.2$ p-value = $P(F(11.9) > 1019.2)$ = 1.54831e-012	$F(11.9) = 16.328$ p-value = $P(F(11.9) > 16.328)$ = 0.000125455	$F(11.9) = 4.83722$ p-value = $P(F(11.9) > 4.83722)$ = 0.0125815
Normality of error distribution test	Chi-squared (2) = 0.00876763 p-value = 0.995626	Chi-squared (2) = 0.246723 p-value = 0.883944	Chi-squared (2) = 1.64938 p-value = 0.43837

*** 1% significance level, ** 5% significance level, * 10% significance level.

Some control variables also appeared to be significant. Firstly, an increase of 1% in the number of women per 1000 men raises the number of employees in a region by 1.23%. It can be explained with the assumption that usually women are paid less than men; therefore they are cheaper for the employer and it raises their employment numbers. Empirically, the lower wages of women were estimated by Arabsheibani and Lau (1999) and Oshchepkov (2006). The positive sign and significance of the variable of life expectancy at birth is also revealing. Its growth over a year increases employment by 1.37%. It reflects the impact of the quality of life in a particular region on the dynamics of the labour force population in the High North. A growth in life expectancy at birth increases the level of regional amenities, and employees can agree to work for lower wages without leaving the northern areas. Increasing the share of people of working age in the general population by 1% leads to a growth in net migration of 3765 people. This fact reflects a concentration of the working age population in areas with higher wages, which attract immigrants. The structure of the regional economy also appears to be significant – a growth in the share of people employed in trade of 1% leads to an increase in net migration of 2337 people. According to the model of Moretti (2011), employment in trade, which refers to the industries that produce so-called 'non-tradable goods' such as services, in the region, is secondary to that in the main industries. In other words, the main industries of the regional economy are the first to develop. They attract more employees, and this affects the development of the service sector, including trade, in order to serve the needs of employees of the other industries. Consequently, the significance and the positive sign of the structure of the regional economy can be explained as a reaction of immigrants not only to the development of the trade itself but also to the growth of employment in the basic industries of the regional economy.

18.3   Conclusions

In general, it can be concluded that the peculiarities of employment in the High North of Russia are based on the specifics of labour supply and labour demand in those regions. A dynamic fixed effects model estimated using the aggregated regional panel data for the High North regions of Russia from 2005 to 2013 demonstrated that wage significantly and positively affects interregional migration to the northern regions. The estimations showed that even in the case where there is a need for additional employees in the High North regions, such a need will be covered only partly by immigrants and partly by the unemployed already living in those regions. This finding indirectly tells us about the surplus of labour supply with respect to labour demand in the High North regions of Russia. The growth of wages attracts immigrants from other regions of the country and eventually leads to lower wages and higher unemployment in the northern territories. It can be assumed that the artificial suppression of emigration from the High North regions of Russia may strengthen these negative consequences.

Acknowledgments   The research was supported by the Russian Foundation for Basic Research # 17-02-00299.

References

Arabsheibani GR, Lau L (1999) "Mind the gap": an analysis of gender wage differentials in Russia. Labour 13:761–774. doi:10.1111/467-9914.00114

Berger M, Blomquist G, Sabirianova PK (2003) Compensating differentials in emerging labour and housing market: estimates of quality of life in Russian cities. Discussion paper 900. Bonn: IZA

Bignebat C (2003) Spatial dispersion of wages in Russia: does transition reduce inequality on regional labour markets? TEAM, University of Paris I &CNRS

Bignebat C (2006) Labour market concentration and migration patterns in Russia. MOISA Working paper 4

Coelho P, Ghali M (1971) The end of the north-south wage differential. Am Econ Rev 61:932–937

Commander S, Nikoloski Z, Plekhanov A (2011) Employment concentration and resource allocation: one-company towns in Russia. Discussion paper 6034. Bonn: IZA

Giltman M (2016) Does location affect employment? Evidence from the high north of Russia. J Urban Reg Anal 8(1):21–36

Gimpelson V, Kapeliushnikov R, Lukiyanova A (2010) Stuck between surplus and shortage: demand for skills in Russian industry. Labour 24:311–332. doi:10.1111/j.1467-9914.2010.00483.x

Greenwood M, Hunt G, Rickman D, Treyz G (1991) Migration, regional equilibrium, and the estimation of compensating differentials. Am Econ Rev 81:1382–1390

Kryukov V, Moe A (2013) Oil industry structure and developments in the resource base: increasing contradictions? In: Godzimirski JM (ed) Russian energy in a changing World: what is the outlook for the hydrocarbons superpower? Ashgate, Farnham, pp 35–56

Mkrtchyan N, Karachurina L (2015) Population change in the regional centres and internal periphery of the regions in Russia, Ukraine and Belarus over the period of 1990–2000s. Bull Geogr. Socio Econ Series 28:91–111. doi:10.1515/bog-2015-0018

Moretti E (2011) Local labour markets. In: Ashenfelter O, Card D (eds) Handbook of labour economics, vol 4B. Elsevier, Amsterdam, pp 1237–1314

Nalimov P, Rudenko D (2015) Socio-economic problems of the Yamal-Nenets Autonomous Okrug development. Procedia Econ Financ 24:543–549. doi:10.1016/S2212-5671(15)00629-2

Nazarova N (2016) Between everything and nothing: organising risks and oil production in the Russian Arctic. Energy Res Soc Sci 16:35–44. doi:10.1016/j.erss.2016.03.024

Oshchepkov AY (2006) Gender wage gap in Russia. HSE Econ J 10(4):590–619

Oshchepkov A (2015) Compensating wage differentials across Russian regions. In: Mussida C, Pastore BK (eds) Geographical labor market imbalances. recent explanations and cures. Iss AIEL series in labour economics. Springer, Berlin, pp 65–105. doi:10.1007/978-3-642-55203-8

Pilyasov AN (2016) Russia's arctic frontier: paradoxes of development. Reg Res Russ 6(3):227–239. doi:10.1134/S2079970516030060

Roback J (1982) Wages, rents and the quality of life. J Polit Econ 90:1257–1278

Rosen S (1979) Wagebased indexes of urban quality of life. In: Miezkowski P, Straszheim M (eds) Current issues in urban economics. Johns Hopkins University Press, Baltimore, pp 74–104

Sardadvar S, Vakulenko E (2016) Interregional migration within Russia and its east-west divide: evidence from spatial panel regressions. Rev Urban Reg Dev Stud 28(2):123–141. doi:10.1111/rurd.12050

Saxinger G (2016) Lured by oil and gas: labour mobility, multi-locality and negotiating normality & extreme in the Russian far north. Ext Ind Soc 3(1):50–59. doi:10.1016/j.exis.2015.12.002

Saxinger G, Öfner E, Shakirova E, Ivanova M, Yakovlev M, Gareyev E (2016) Ready to go! The next generation of mobile highly skilled workforce in the Russian petroleum industry. Ex Ind Soc 3(3):627–639. doi:10.1016/j.exis.2016.06.005

Vakulenko E (2016) Does migration lead to regional convergence in Russia? Int J Econ Policy
    Emerg Econ (IJEPEE) 9(1):1–25. doi:10.1504/IJEPEE.2016.074943
World Bank (2010) Implementation completion and results report (IBRD-46110) on a loan in the
    amount of US$ 80 million to the Russian Federation for a Northern restructuring project. The
    World Bank Report No: ICR00001343, June, 21. Available at: http://www-wds.worldbank.
    org/external/default/WDSContentServer/WDSP/IB/2010/07/19/000333037_201007190005
    42/Rendered/PDF/ICR13430ICR0Bo1B01Official0Use01091.pdf (date Accessed 09/27/2016)

Chapter 19
Well-Being in an Arctic City. Designing a Longitudinal Study on Student Relationships and Perceived Quality of Life

John A. Rønning, Steinar Thorvaldsen, and Gunstein Egeberg

Abstract Research on bullying and harassment in Scandinavia has been going on for several decades, and is appearing in new frameworks and forms since the new categories of "cyber-harassment" or "cyber-bullying" have been introduced. Bullying is a phenomenon of great worry, as it seems to affect children and adolescents both on short and long term. A questionnaire on cyber-harassment was designed in this study, and answered by pupils and their parents and teachers, at five schools in the city of Tromsø, Norway. The questionnaire included a section of questions concerning traditional forms of harassment and bullying, as well as questions on quality of life (QoL), and the Strength and Difficulties Questionnaire (SDQ). The main research questions were: (1) What is the prevalence of the three classical types of bullying and cyberbullying; (2) Are there gender or age differences; (3) What percentage of children bullied classically were also cyberbullied; (4) How and to what extend did those that were bullied also suffer a lower quality of life. The main novel contribution of this study to the ongoing research is that students who reported being cyber-harassed or cyberbullied, also reported significantly lower QoL-scores than their non-harassed peers.

19.1 Introduction

It is widely acknowledged that children's experience on transactions with peers play an important part in children's development and socialization (Hartup 1978; Harris 1998; Scarr 1992). Perceived positive relations with others are associated with the enhancement of social understanding (Dunn 1999) and social competency and positive adaptation in childhood and in school (Hartup 1983), but also with positive adaptation later on in life (Hartup 1976; Parker and Asher 1987). Parker and Asher (1987) found evidence of the fact that children experiencing poor adjustment with

J.A. Rønning • S. Thorvaldsen (✉) • G. Egeberg
UIT The Arctic University of Norway, Tromsø, Norway
e-mail: steinar.thorvaldsen@uit.no

© The Author(s) 2017   185
K. Latola, H. Savela (eds.), *The Interconnected Arctic — UArctic Congress 2016*,
Springer Polar Sciences, DOI 10.1007/978-3-319-57532-2_19

others were at high risk of developing serious problems in adulthood. Thus, experienced relationships to others in school are of fundamental importance for positive or poor adaptation later on in life.

19.1.1   Bullying and Harassment

Bullying of other children or being a victim of bullying have repeatedly been documented to exert negative influence on children's development with serious long-term effects. In a large comparative study on the prevalence of bullying in various western countries involving 123,227 students aged 11, 13 and 15 uncovered the following: the lowest prevalence of bullying or being bullied was reported by Swedish girls (6.3%), while the highest came from boys in Lithuania (41.4%). Generally, girls were less involved in bullying compared with boys (Due et al. 2005). In a study from northern Norway involving 4167 girls and boys from 66 schools, about 5% reported being bullied (Rønning et al. 2004). Most of these were boys.

Olweus (1999) defines bullying as a situation when a child/student repeatedly experience negative reactions over time from one or more fellow students who intentionally apply these reactions and where the victim cannot defend him-/herself. Bullying is often divided into four categories: (i) direct/physical (physical attack or theft); (ii) direct verbal (threats, insults, calling names); (iii) indirect/relational (social exclusion, spreading of nasty rumors) and (iv) cyber bullying (text messages, posting pictures, spreading rumors, exclusion from social media like Facebook). Cyber bullying is a new phenomenon. Thus, knowledge of its prevalence and short and long-term effects is scarce. However, research on cyber bullying is increasing (Slonje and Smith 2008; Smith et al. 2006). Most studies report its prevalence and correlate to classical bullying. In this project we will report on all types of bullying and also relate them to the perceived quality of life and mental health.

Research (Frisén et al. 2008; Rønning et al. 2009; Smith 2002; Ybarra et al. 2012) has uncovered uncertainty about the definition of bullying and there is a disagreement between parents, teachers and children about its prevalence, and thus it's prevention. With the introduction of cyber bullying, Slonje and Smith (2008) argue for a debate on the criteria for something to be called bullying. The debate is especially focused on the criteria for repetition. As an example, they discuss the event of posting of unpleasant pictures on the web. Even though only one picture is posted, many might see it, and it may stay on the web for a long time. Thus, it seems imperative to continue to conduct both qualitative and quantitative studies on how students, teachers and parents define bullying, improve the measurement of bullying and study both short and long-term effects of bullying on well-being.

19.1.2   Mental Health

To experience being bullied over a long time is considered to be one of the most stressful life events (Branwhite 1994), and those experiencing this are placed in a high-risk position regarding the development of a negative self-image and poor adaptation (Deković and Gerris 1994). It has also been uncovered that children bullying other children are more at risk of later psychiatric problems and criminality than their victims are (Kumpulainen and Raesaenen 2000; Sourander et al. 2007a, b). For example, Sourander et al. (2007a, b) uncovered that 28% of the boys in a representative sample of boys born in 1981 who were reported being bullies or bully-victims in the age of 8 years, had a psychiatric diagnosis 10 years later. Altogether 33% of these boys were found in the Finnish criminal registry when they were 16–20 years of age. However, this was relevant only for those boys reporting psychiatric symptoms when they were 8 years old. Thus, an early combination of bullying behavior and psychiatric problems is a very strong predictor/factor indicating later psychiatric disorders and criminality. A study in Norway (Rønning et al. 2004) found strong associations between being a victim and problems with friends and behavior problems. Another study (Salmon et al. 2000) found that victims were referred to mental health services due to depression and generalized anxiety, whereas bullies were referred because of behavior problems and ADHD.

Several studies have shown associations between problematic child – child relationships and later criminality (Roff et al. 1972), school refusal (Parker and Asher 1987), military records with serious behavior problems (Roff 1961), manic-depressive and schizophrenic disorders (Cowen et al. 1973; Kohn and Clausen 1955) and suicide (Stengel 1971). These problems have also been associated with parental problems like poverty, substance abuse, psychiatric problems, and child neuropsychological problems. All studies mentioned here have focused on the relationship between mental health and classical bullying. So far, we know very little about the relationship between cyber bullying and mental health.

19.1.3   The School Culture

The bystanders of bullying may have an essential influence on bullying. It has been uncovered that other students have been involved in 85% of bullying episodes, either as observers or as direct participants (Craig and Pepler 1995, 1997). In particular Salmivalli and associates (2004) have documented that bullying is reinforced by fellow students. This may lead to such behavior being regarded as acceptable and normative within the peer group. It is speculated that the main reason for the poor effect of anti-bullying programs is lack of understanding of the importance of the school culture (Swearer et al. 2009).

19.1.4 Quality of Life

Both bullying and victimization are associated with the experience of poor quality of life (Thorvaldsen et al. 2016). Quality of life is the individual's experience of life being satisfactory. Quality of life is a multidimensional concept, which includes physical and emotional well-being, self-image, relationships within family and amongst friends and the daily functioning in school. Jozefiak et al. (2008) investigated the perceived quality of life (KINDL-r) of 1997 randomly selected students aged 8–16 in the middle of Norway (participation 71.2%) The study uncovered acceptable psychometric properties of the instrument; the children scored themselves lower than the parents, and girls lower than boys. This study will constitute a reference in the present study.

19.1.5 Research Questions

The main research questions in this study were: What is the prevalence of various types of bullying, and what are the associations to mental health and quality of life? Examples of various questions to be answered are the following:

1. What are the percentage of bullies, bully-victims, victims and bystanders?
2. Are those related to cyber bullying the same as those related to classical bullying?
3. What characterizes the various bullying types?
4. What is the relationship between the students' well-being and their functioning in school?
5. How is the mental health of students in the study?
6. How do the students in the study perceive their quality of life?
7. How do bullying influence the mental health and adaptation in school?
8. Are there changes in the pattern of experienced peer relationships and mental health and quality of life across time?

In addition to these quantitative problems, it was of interest to collect qualitative data in order to better understand the school culture. Because the teacher education in Tromsø has been integrated into the University of Tromsø and became a Master's degree program, involvement of students in answering research questions is an integrated aim of the project.

19.2 Methods

19.2.1 Design

The study was designed both as a prospective longitudinal and as a cross-sectional time trend study starting in late fall 2013. In the prospective longitudinal part of the study, the same students will be followed for a period of up to 6 years. The long term study is still ongoing, as shown in Fig. 19.1.

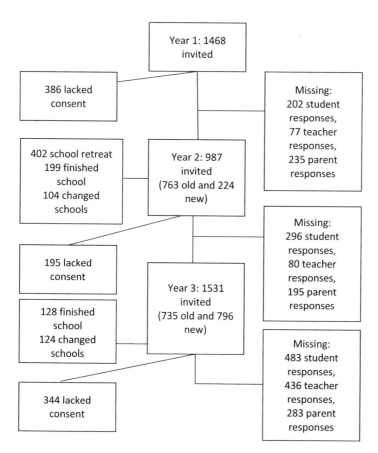

Fig. 19.1 Flow chart showing participation and dropouts for the Well-being in Tromsø survey in the first 3 years of the main project. There are overlap regarding missing data and lack of consent. The survey is an ongoing project that started in the school year 2013/2014

19.2.2   Participants

A pilot project was carried out at one of the teacher training schools in order to establish the most effective logistics. The aim of the formal project was to include five teacher training schools (grade 4–10) a total number of ~1000 children/students (50% girls). These children will be followed up with the same questionnaires every year for a period of 6 years as a time trend study. The child/student, teacher and parents constitute the informants. On the perceived quality of life questionnaire (KINDLE-r) only the student and parents constitute the informants. It is of uttermost importance that the research project connects the researcher, student, teacher and parents and thus promotes the research and research-based knowledge in the education of teachers.

19.2.3   Instruments

Demographics A questionnaire asking for gender, about parents' occupations, and how many years of education they have was applied.

Classical Harassment Classical harassment was divided into three categories: "physical aggression" – 4 questions; "verbal harassment" – 5 questions; and "social manipulation" – 6 questions. These categories and questions are derived from the study of Rønning et al. (2004). The instrument demonstrated very good psychometric properties.

Cyber Harassment This part of the investigation builds on questionnaires developed by Smith et al. (2006) and Menesini el al. (2011). In the investigation it was asked how often participants have experienced cyber bullying during the last 2–3 months in the different areas of mobile phone and internet. The first questions are general about prevalence when they attended school and outside school, and to what degree the student had participated in the bullying. It was followed up with a series of questions covering ten types of cyber bullying (SMS, MMS, phone calls, e-mail, internet text, instant messages, chat, blogs, internet video, and exclusion from social media like Facebook).

A five point Likert scale was applied on all types of bullying (never, only once or twice, two or three times a month, about once a week, many times per week). It was also asked what reactions they had received when bullying is alerted, and to what degree the students themselves and the teacher tried to stop it.

Strengths and Difficulties Questionnaire (SDQ) SDQ is a screening instrument for behavioral and emotional problems that consists of 25 questions distributed equally on the following dimensions: "emotional symptoms", "behavior problems", "hyperactivity", "problems with friends" and "pro-social behavior". A "total problem score" is calculated (Goodman 1997). The SDQ comes in different versions for children of different age, teachers and parents.

Quality of Life (KINDL-r) KINDL-r measures perceived quality of life (Ravens-Siberer and Bullinger 2000). There were two versions; one where the parents evaluate the child, and one where the child evaluates own quality of life. Every version consists of 24 questions. Each question asked for the last week experiences, and was scored on a 5-point scale ("never", "rarely", "sometimes", "often", "always"). The psychometric properties of KINDL-r are considered as very good (Jozefiak et al. 2008), but for the Norwegian version more validation work is requested.

19.2.4 Ethics

The study was carried out digitally using the commercial online survey tool "Questback". Only the project leadership of the University of Tromsø had access to the filled-in questionnaires. For student's inclusion to the study, the parents must have given their signed informed consent. Students and parents were able to resign at any time from the study without grounds, and data that have not already been published will be deleted. A separate consent regarding those participating in the interview part of the study will be obtained.

Feedback to schools was submitted through annual dialog conferences, and each school received a report where it can be compared with the gross mean. If wanted, each school will also receive a report where data from each class was published. The project was approved by the Regional Ethical Committee for Medical Research (REK-Nord).

19.3 First Results from the Study

Pilot Study The design, administration and the questionnaire of the project was tested out at one pilot school in 2012/2013. Analyses from the pilot study revealed several interesting results. There were differences on how students reported being bullied. Low occurrence was reported when asked how often they are being bullied, but at the same time, the students report higher frequencies of concrete harassment actions. In other words, they reported that they were not being bullied, but still they were quite often exposed to name-calling, teasing and various forms of negative physical actions. In addition, students at grade 7–9 reported significantly *lower* bullying than at younger grade (4–6). Monks and Smith (2006) point out that age is a factor how they define bullying. Older children (age 14) normally demonstrate a more differentiated understanding while the younger ones (age 6–8) often relate bullying to physical aggression. This is in line with other reports on this issue. Boulton et al. (2002) find that children do not necessarily regard social exclusion, name-calling and stealing as bullying, whereas the physical categories are more often defined as bullying by the children.

Long-Term Study Some results from the first year of the study are recently reported in Thorvaldsen et al. (2016), and in Egeberg et al. (2016). It was found that, compared to traditional bullying, cyberbullying is less common, but it now affects some 3.5% of pupils, i.e. more than one third of the level of classical bullying, and most of it takes place outside of school. There are only small gender-related differences in incidence at both primary school (ages 10–13) and secondary school (ages 14–16), contrary to many earlier research reports.

Both "traditional" and cyber forms of bullying and harassment show significantly ($p < 0.05$) lower scores on their self-perceived quality of life factors. Non-victims reported a mean between 4.1 and 3.4 on a scale from 1 to 5, while those who reported having been bullied reported a mean between 3.7 and 3.0. Cyber-harassment and cyber-bullying share the same negative characteristics in relation to quality of life as classical harassment and bullying. Using Structural equation modeling (SEM) analyses, it was found that both cyber harassment and cyber bullying had a distinct and substantial impact on students' academic achievements. However, this effect was largely mediated through a reduction in students' perceptions of quality of life. Thus, it is important to address the issue of perceived quality of life, especially for those students being subjects to bullying and severe harassment.

To elaborate more about these issues and on how students perceive the term "bullying", focus group interviews with students and with teachers were conducted. In the various interviews, the general term and the specific categories, and age variations (Egeberg et al. in press) of bullying were examined. Furthermore, the issues of perceived severity in specific negative conduct and point at differences between teachers and students were addressed in this regard.

19.4   Concluding Remarks

In order to determine whether the results obtained from this study are stable over time, and to produce a more detailed study of causalities between the variables, a longitudinal study is necessary, and this project will accomplish that in the near future (Fig. 19.1). The results obtained from long-term studies like this one may lead to a deeper understanding of student relations, and the development of much-needed policy and methods of preventing and intervening in cases of harassment and bullying.

References

Boulton MJ, Trueman M, Flemington I (2002) Associations between secondary school pupils' definitions of bullying, attitudes towards bullying and tendencies to engage in bullying: age and sex differences. Educ Stud 28(4):353–370
Branwhite T (1994) Bullying and student distress: beneath the tip of the iceberg. Educ Psychol 14(1):59–71

Cowen EL, Pederson A, Babigian H, Izzo LD, Trost MA (1973) Long-term follow-up of early detected vulnerable children. J Consult Clin Psychol 41:438–446

Craig WM, Pepler DJ (1995) Peer processes in bullying and victimization. An observational study. Except Educ Can 5:81–95

Craig WM, Pepler DJ (1997) Observations of bullying and victimization in the school yard. Can J Sch Psychol 13:41–59

Deković M, Gerris JRM (1994) Developmental analysis of social cognitive and behavioral differences between popular and rejected children. J Appl Dev Psychol 15:367–386

Due P, Holstein BE, Lynch J et al (2005) The health behaviour in school-aged children bully working group. Bullying and symptoms among school-aged children: international comparative cross sectional study in 28 countries. Eur J Pub Health 15:128–132

Dunn J (1999) Siblings, friends, and the development of social understanding. In: Collins AW, Lauritsen B (eds) Relationships as developmental contexts. The Minnesota symposia on child psychology, vol 30. Lawrence Erlbaum Associates, New York, pp 263–279

Egeberg G, Thorvaldsen S, Rønning JA (2016) The impact of cyberbullying and cyber harassment on academic achievement. In: Digital expectations and experiences in education. Sense Publishers, Rotterdam, pp 183–204

Egeberg G, Thorvaldsen S, Rønning JA (Manuscript in press) Understanding bullying: how students and their teachers perceive terms of negative conduct

Frisén A, Holmqvist K, Oscarsson D (2008) 13-year-olds' perception of bullying: definitions, reasons for victimisation and experience of adults' response. Educ Stud 34(2):105–117

Goodman R (1997) The strenghts and difficulties questionnaire: a research note. J Child Psychol Psychiatry 38:581–586

Hartup WW (1976) Peer interaction and behavioral development of the individual child. In: Schopler E, Reichler RJ (eds) Psychopathology and child development. Plenum Press, New York

Hartup WW (1978) Children and their friends. In: McGurk H (ed) Issues in childhood social development. Methuen, London, pp 130–170

Hartup WW (1983) Peer relations. In: Mussen P (series ed) & Hetherington E (vol ed), Handbook of Child Psychology, (vol 4): Socialization, personality and social development. Wiley, New York, pp 103–198

Harris JR (1998) The nurture assumption. Why children turn out the way they do. Parents matter less than you think and peers matter more. The Free Press, New York

Jozefiak T, Larsson B, Wichstrøm L, Mattejat F, Ravens-Sieber U (2008) Quality of life as reported by school children and their parents: a cross-sectional survey. Health Qual Life Outcomes 6:34

Kohn M, Clausen J (1955) Social isolation and schizophrenia. Am Sociol Rev 20:265–273

Kumpulainen K, Raesaenen E (2000) Children involved in bullying at elementary school age: their psychiatric symptoms and deviance in adolescence. An epidemiological sample. Child Abuse Negl 24:1567–1577

Menesini E, Nocentini A, Calussi P (2011) The measurement of cyberbullying: dimensional structure and relative item severity and discrimination. Cyberpsychol Behav Soc Netw 14(5):267–274

Monks CP, Smith PK (2006) Definitions of 'bullying'; age differences in understanding of the term, and the role of experience. Br J Dev Psychol 24:801–821

Olweus, D. (1999). In: Smith PK, Morita Y, Junger-Tas J, Olweus D, Catalana R, & Slee P. (eds) The nature of school bullying. A cross-national perspective. Routledge, New York

Parker JG, Asher SR (1987) Peer relations and later personal adjustment: are low-accepted children at risk? Psychol Bull 102:357–389

Ravens-Siberer U, Bullinger M (2000) KINDL-R Questionnaire for measuring health-related quality of life in children and adolescents – revised version 2000 (http://www.kindl.org)

Roff M (1961) Childhood social interactions and adult bad conduct. J Abnorm Soc Psychol 63:333–374

Roff M, Sells B, Golden MM (1972) Social adjustment and personality development in children. University of Minnesota Press, Minneapolis

Rønning JA, Handegård BH, Sourander A (2004) Self-perceived peer harassment in a community of Norwegian school children. Child Abuse Negl 28(10):1067–1079

Rønning JA, Sourander A, Kumpulainen K et al (2009) Cross-informant agreement about bullying and victimization among eight-year-olds: whose information best predicts psychiatric caseness 10–15 years later? Soc Psychiatry Psychiatr Epidemiol 44:15–22

Salmivalli C, Voeten M (2004) Connections between attitudes, group norms and behavior associated with bullying in schools. Int J Behav Dev 28:246–258

Scarr S (1992) Developmental theories for the 1990's: development and individual differences. Child Dev 63:1–19

Salmon G, James A, Cassidy EL, Javaloyes AM (2000) Bullying a review: presentations to an adolescent psychiatric service and within a school for emotionally and behaviorally disturbed children. Clin Child Psychol Psychiatry 5:563–579

Slonje R, Smith PK (2008) Cyberbullying: another main type of bullying? Scand J Psychol 49(2):147–154

Smith PKP (2002) Definitions of bullying: a comparison of terms used, and age and gender differences, in a fourteen-country international comparison. Child Dev 73(4):1119–1133

Smith PK, Mahdavi J, Carvalho M, Tippett N (2006) An investigation into cyberbullying, its forms, awareness and impact, and the relationship between age and gender in cyberbullying, Research brief no RBX03-06. DfES, London

Sourander A, Jensen P, Rønning JA et al (2007a) What is the early adulthood outcome of boys who bully or are bullied in childhood? The Finnish "From a boy to a man" study. Pediatrics 120:397–404

Sourander A, Jensen P, Rønning JA et al (2007b) Are childhood bullies and victims at risk of criminality in late adolescence? The Finnish "From a boy to a man" study. Arch Pediatr Adolesc Med 161:546–552

Stengel E (1971) Suicide and attempted suicide. Penguin, Middelsex

Swearer SM, Espelage DL, Napolitano SA (2009) Bullying prevention and intervention: realistic strategies for schools. Guilford, New York

Thorvaldsen S, Stenseth AM, Egeberg G, Pettersen GO, Rønning JA (2016) Cyber harassment and quality of life. In: Digital expectations and experiences in education. Sense Publishers, Rotterdam, pp 161–182

Ybarra ML, Boyd D, Korchmaros JD, Oppenheim JK (2012) Defining and measuring cyberbullying within the larger context of bullying victimiza-tion. J Adolesc Health 51(1):53–58. doi:10.1016/j.jadohealth.2011.12.031

Chapter 20
Researching Links Between Teacher Wellbeing and Educational Change: Case Studies from Kazakhstan and Sakha Republic

Olga M. Chorosova and Nikolai F. Artemev

Abstract Many issues about teacher evaluation have been discussed in Russia where evaluation procedures are constantly under review as the teachers seek continuous improvement. This chapter contributes to this discussion, first, by adding a wider international perspective, secondly, by exploring the experiences of participants (observers, in-service teachers) and, finally, by drawing on research related to teacher evaluation. The authors present the outcomes of the survey conducted among teachers in Russia's Sakha Republic (Yakutia) and in the Republic of Kazakhstan. The research objectives were: evaluation of the adaptation process in professional and personal development of in-service teachers caused by changes in education system of Russia and the development of basic professional competences of teachers. The study aimed to explore the views of teachers and analyses of teacher evaluation via a mixed-method approach. The rationale for this study was born out of the disillusionment with teacher evaluation. Therefore, it was concluded that existing evaluation practices may need to be examined to see if they serve the best interests of teachers.

20.1 Introduction

Yakutia and Kazakhstan share common Soviet past and have presently similar state of education (Chorosova et al. 2006). At different stages of its development, Russian educational system has experienced processes of modernization; at different times attempts were made to reform it so that at this stage we can talk about systemic changes which are understood as a transformation of the educational space. Excerpts from the Professional Standard show conclusively the relationship between the moral side of a teacher's professional activity and emotional-volitional part of his or her personality as it is only a teacher who is capable of spiritual and moral reflection

O.M. Chorosova (✉) • N.F. Artemev
Ammosov North-Eastern Federal University,
58 Belinsky St, Yakutsk 677000, Russian Federation
e-mail: chorosovaom@mail.ru; nf.artemev@s-vfu.ru

© The Author(s) 2017
K. Latola, H. Savela (eds.), *The Interconnected Arctic — UArctic Congress 2016*,
Springer Polar Sciences, DOI 10.1007/978-3-319-57532-2_20

and widely accepting continuous self-improvement and lifelong education. A teacher who can reveal to students strategies of life and tactics of self-improvement, and provide professional and pedagogical help in designing their personal development (Asmolov 1996). At the basis of this activity is the ability of teachers to spiritually and morally improve their personality, which has a limitless potential by such cognitive indicators as: (1) the volume of knowledge on morality, its connection to spirit and spirituality; (2) consistency of this knowledge; (3) its meaningfulness; (4) its stability; (5) its creative use in life. All of the above makes the issue of basic professional competencies of a teacher extremely relevant (Amabile 1982). In the beginning of 2000s, the Russian society started to overcome the negative tendencies of the crisis of the end of the 1990s. The pedagogical society was experiencing an emotional uplift, the feeling of a certain stability in social and economic context began to dominate. The beginning of 2000s was characterised by an unprecedented innovative boom in the education of the republic, which started actively integrate into the international open educational environment. This probably accounts for the highest value of the openness to the world factor in 2005 (Chorosova et al. 2006).

20.2 Problem Statement

The literature on teachers' evaluations were carefully studied. One of the objectives of this study was to understand how teachers are evaluated and supported throughout their careers but there were some consequent problems (Howard and Donaghue 2015). Before describing how the study was conducted, an overview is given to pertinent themes emanating from literature on this area of concern.

The main problem is a rise in managerialism in education that leads to "extra scrutiny" since people want proof of teachers' effectiveness (Deem 2003). Deem (2003) suggests a more negative perception which focuses on functions, tasks and behaviors. This has institutions to become linemanaged entities where 'professionals are subjected to a rigorous regime of external accountability in which continuous monitoring and audit of performance and quality are dominant' (Deem 2003, pp. 57–58). Kydd (1997) summarizes the conflict effectively by suggesting that 'the intensification of management controls is replacing the wisdom, experience, and self-monitoring of the practitioner' (p. 116).

The focus was on problems related to evaluation of in-service teachers. While such monitoring may be familiar in pre-service teacher training, for experienced teachers it may be more stressful (Howard and McCloskey 2001) as they might feel that the survey questions interfere with their professionalism and that is inappropriate. According to Howard and McCloskey, major reasons for an aversion to evaluation are the sense that it does not promote professional growth and its accountability and prescriptive conformity which may conflict with a teacher's desire for professional autonomy (Fullan 2007). Quality assurance (QA) is defined by its adherence to measurable standards and outcomes. An education has become an international enterprise, and the need to ensure that trans-border standards are met has magnified its importance (Aubrey and Coombe 2011). QA obliges practitioners to focus on teaching curricula efficiently and effectively (Kydd 1997). However, there are certain precautions to consider its successful implementation.

This study aimed to explore, via a mixed-method approach, the views of experienced teachers and analysts on teacher evaluation. The rationale for this study was born out of our own disillusionment with teacher evaluation. Therefore, the existing evaluation practices may need to be examined to see if they serve the best interests of experienced teachers. To achieve this, following research questions were designed: What is the quality of professional and personal wellbeing of teachers? How to diagnose and prevent reasons of emotional burnout of teachers? How to define ways of reducing crisis factors' influence on northern teacher staff's mass exodus? What are the views of experienced teachers regarding teacher evaluation? What are the views of analysts regarding teacher evaluation? The answers to these questions allowed to find a scientific understanding of staff turnover and to define ways of reducing crisis factors' influence on northern teachers and on teachers' emotional burnout. The study of the role of renewal processes and crisis factors of professional and personal wellbeing of teachers, revealing the qualification gaps of teachers through the assessment of professional competences and abilities, will improve teachers' assessment of the quality of life. The strong belief was that the sources of motivation in the professional and personal development of teachers, integrating in creative activities to create innovative projects were curiosity and interest, setting learning goals and belief in self-efficacy. Specifying the structure of creative teaching activities, V.A. Kan-Kalik and N.D. Nikandrov defined the following sequence of its stages: the emergence of the pedagogical plan aimed at addressing educational problems; concept development; the embodiment of the pedagogical design in activities, in dealing with people; analysis and evaluation of the results of the creative process (Kan-Kalik and Nikandrov 1990). The concept of "creativity" was used by S.L. Rubinstein as "contagion" – as creation of new, original case with social significance; as the creation of something new, including in the inner world of the subject (Rubinstein 2006). The environment, in which the Northern teacher can develop creatively, should have a high degree of uncertainty and potential multi-variance (wealth of opportunities). Uncertainty encouraged the search for persons own benchmarks rather than the adoption of what's ready; multi-variance made it possible to find them.

20.3 Methods

A quantitative analysis was required to identify how current teachers overcome obstacles and challenges created by constant stream of changes in the national system of education both in Russian Yakutia and in Kazakhstan. In this study, surveys of 2005, 2015, and 2016 were used. An online questionnaire was developed on the university website platform. The goal was to reveal how teachers face negativity of crisis factors and find ways to continue personal professional development.

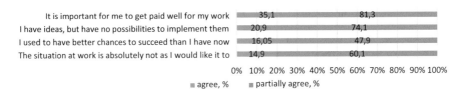

Fig. 20.1 Results of Yakutia survey

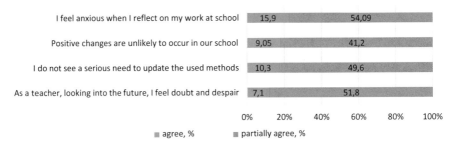

Fig. 20.2 Results of Yakutia survey

20.4 Results

In total, over 1000 in-service teachers of the Sakha Republic (Yakutia) and 2389 in-service teachers of Kazakhstan were surveyed. Both results were compared and analyzed. Teachers of Yakutia (Figs. 20.1 and 20.2) describe the presence of crisis-producing factors. Individuals in a crisis tend to have higher values for these factors. Such individuals are alienated from others and believe that nobody understand them well that external circumstances dictate their life and that they are not able to determine their own lives. They cannot see the meaning of their life, their inner hierarchy of values disintegrated, things that seemed important on a regular basis changed and they cannot decide what is important and what is not. Such individuals are no longer interested in something new and do not seek to develop themselves. It is important to pay attention to the rate of values, which have the highest points relatively to other factors.

Figures 20.3 and 20.4 describe the presence of factors that help to overcome crises in Yakutia survey results. Individuals in a stable condition overcome a crisis and tend to have high values for these factors. Individuals who overcome the crisis are open to people, to the world and they care about not only their own problems but also problems of others and seek to resolve them.

Such individuals constantly educate themselves (Fig. 20.4), work on personal development and are interested in new information. All this allows them to better understand themselves and to set clear goals, make decisions and follow them. Such individuals acquire self-control; they overcome external circumstances and use their

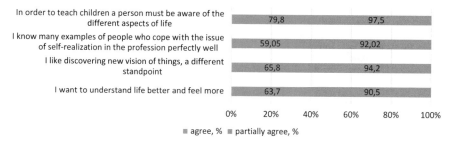

Fig. 20.3 Factor of orientation towards something new, intent for development (Yakutia)

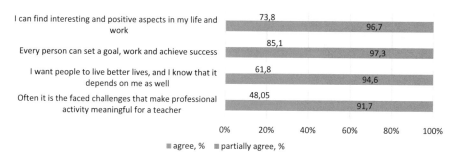

Fig. 20.4 Factor of setting goals (Yakutia)

capacities. It is important to pay attention to the rate of values, which have the lowest points relatively to other factors of overcoming crises. Figures show the representation of a crisis-prone factors in the lives of teachers, clearly show the state of depression, confusion and anxiety in a greater or lesser extent. For instance, 41.2% to 54.09% of Yakutian teachers have the lack of dignity in professional career. 47.9% to 81.3% of teachers are affected by the influence of external negative circumstances, the inability to implement their plans and prioritize high payment for their work experience. 32.1% to 67.4% of teachers have the sense of loss of meaning in life. 28.1% to 60.8% have the feeling of loneliness and alienation. This suggests that teachers respond to unfavorable trends of our times as sensitive indicators of the state of society, primarily due to the fact that this state of society is reflected and refracted in children. However at the same time, the outcomes of this survey showed a very high resistance (from 90.5% to 97.5%) to the effects of a crisis-prone factors accompanied by exceptional adaptability to the adverse social factors and the adaptability to the rapidly changing world and society when there is a need to react quickly and learn fast. Teachers of the North are characterized with a desire to everything new and tolerant attitude, self-control when necessary and self-regulation. 72.6% to 95.7% responses confirm it. Yakutian teachers are aware of the importance of goal-setting skills, when setting more and more new purposes becomes an internal necessity of constantly developing personality: 91.7%–97.3%. A developing personality can be curious, indifferent, open to everything new and unknown at every moment of life. For the analysis of Yakutia and Kazakhstan, it is

Table 20.1 The effect of crisis factors on the teachers of Kazakhstan

A	B	C	D	E	F	G	H
Lack of future prospects	External circums- tances	Lose of reasons for living	alienation	Desire to new, intense to self-development	Gaining self- control	gaining goals	Openness towards the world
0.42	0.45	0.37	0.35	0.7	0.6	0.77	0.75

0 – negative answer, 1 – affirmative answer

Table 20.2 Sakha Republic (Yakutia), 2005

A	B	C	D	E	F	G	H
Lack of future prospects	External circums- tances	Lose of reasons for living	alienation	Desire to new, intense to self-development	Gaining self- control	gaining goals	Openness towards the world
0.3	0.5	0.6	0.6	0.8	0.8	0.8	0.9

0 – negative answer, 1 – affirmative answer

necessary to clarify that selected identical corresponding questions were selected in materials of Kazakhstan colleagues. In identifying crisis factors, it was important to focus on the representation of those of them whose values are at a maximum compared to other factors. During the data analysis, it showed that for Kazakhstan teachers a potential crisis state can be triggered by the influence of external circumstances (Zh. O. Zhilbaev. National Academy of Education). In the Kazakhstan community of teachers, the choice of a behavioural pattern not only in a professional sphere but also in a personal life is often influenced by public opinion. The outcomes revealed that the indicators of all four crisis factors have low values, which show that there is no crisis state; the other three factors (lack of future prospects, losing one's reasons for living, alienation) that represent a serious psychological problem and are potentially dangerous for the society are minimal (0.4–0.5 points). In identifying factors responsible for overcoming crises, it was important to pay a special attention to those of them, whose values are at a maximum compared to other factors. There are two such factors: gaining goals (Table 20.1, column G) and openness towards the world (Table 20.1, column H).

For Kazakhstan teachers, factors responsible for overcoming crises are gaining new goals and being open to the world. Overall, such situation can also be considered as a positive tendency. For the teachers of Yakutia, among the factors evidencing the presence of crisis phenomena the highest value in 2005 was shown by the factor of losing one's reason for living (Table 20.2), and in 2015 the factor of external circumstances (Table 20.3) even though the average value was 0.3 to 0.6 in 2005 and 0.27 to 0.45 in 2015–2016.

Looking at these values, we can see that in 10 years the teachers of Yakutia become less susceptible to the influence of crisis factors. In 2015, the highest values are shown by the factors of gaining new goals and openness to the world as among

Table 20.3 Sakha Republic (Yakutia), 2015–2016

A	B	C	D	E	F	G	H
Lack of future prospects	External circums-tances	Lose of reasons for living	alienation	Desire to new, intense to self-development	Gaining self-control	gaining goals	Openness towards the world
0.27	0.45	0.35	0.3	0.7	0.6	0.8	0.8

0 – negative answer, 1 – affirmative answer

the teachers of Kazakhstan. The results show that teachers are in a stable condition: they are open towards other people and the world, they are concerned not only with their own problems but also with problems of others, their students and society, the teachers strive to resolve those problems and they feel that they belong to the society. The teachers cultivate their own abilities, change and strive to everything new.

All of the above allows them to be self-aware, to set clear goals, to make decisions and be responsible for them. Therefore, in order to neutralise the crises in professional development, the greatest significance is given to the factors of gaining goals and being open to the world.

20.5   Conclusions

One of the goals of the continuous professional education is the creation of platforms where teachers find the opportunity to discuss their professional work from the value-semantic content point. It is paramount for the education as the teacher who has not found personal way in understanding the mission of teaching and who is careless and blind about meaning of teaching cannot educate and bring up children, that teacher cannot nourish moral and ethical foundations. The survey aimed to identify the formation of professional competences in the context of the current requirements, compliance with professional standards for teachers, reveal qualification deficits in teachers of the Sakha Republic (Yakutia) and Kazakhstan, both urban and rural, and they, in turn, determine the educational needs of teachers in continuous professional education– training courses or retraining. From a personal perspective, the findings have shown that the questioning of the value of evaluation is shared to a certain extent in fact there are signs of discontentment among the population. However, there is also an apparent belief in the purpose of evaluation and if it is designed and implemented in a way, which motivates teachers and helps them to become better practitioners. Therefore, there is a need to forego opinions, knowing that teachers often find it a worthwhile activity, as ultimately staff improvement leads to better teaching and learning and this shows higher education quality. While teacher evaluation may have negative connotations for some, quality does not, so if everyone will aim for quality in teaching and learning, it certainly would seem a worthwhile endeavor.

References

Amabile TM (1982) Social psychology of creativity: a consensual assessment technique. J Pers
    Soc Psychol 43(5):997–1013
Asmolov AG (1996) Cultural-historical psychology and design worlds. Institute of Practical
    Psychology, Moscow; MODEK, Voronezh, 768
Aubrey J, Coombe C (2011) An investigation of occupational stressors and coping strategies
    among EFL teachers in the United Arab Emirates. In: Gitsaki C (ed) Teaching and learning in
    the Arab world. Bern, P. Lang, pp 181–201
Chorosova OM, Gerasimova RE, Nikiforov ES, Alexeyeva FI (2006) A teacher in a changing
    society. 104
Deem R (2003) New managerialism in UK universities: manager-academic accounts of change.
    In: Eggins H (ed) Globalisation and change in higher education. The Society for Research into
    Higher Education (SRHE), Open University Press, Berkshire
Fullan M (2007) The new meaning of educational change, 4th edn. Teachers College Press,
    New York
Howard A, Donaghue H (eds) (2015) Teacher evaluation in second language education. London,
    Bloomsbury. 220
Howard BH, McCloskey WH (2001) Evaluating experienced teachers. Educ Leadersh 58(5):48–51
Kan-Kalik VA, Nikandrov ND (1990) Teacher creativity. M.: Pedagogika, 140. ISBN 5-7155-0293-4
Kydd L (1997) Teacher professionalism and managerialism. In: Kydd L, Crawford M, Riches
    C (eds) Professional development for educational management. Open University Press,
    Buckingham, pp 111–117
Kydd L, Crawford M, Riches C (1997) Professional development for educational management.
    Open University Press, Buckingham, pp 111–117
Rubinstein SL (2006) Fundamentals of general psychology. School Manag 3:27–29

Chapter 21
Well-Being at the Polish Polar Station, Svalbard: Adaptation to Extreme Environments

Anna G.M. Temp, Billy Lee, and Thomas H. Bak

Abstract While the psychological well-being of Antarctic crews has been investigated previously, Arctic crews have received little attention. Antarctic stressors include the permanent darkness of polar night, cramped quarters and harsh weather conditions which demand that the crews work together to survive. These stressors are also present for Arctic crews with the addition of dangerous polar bears. In this study, these psychological stressors were explored at the Polish Polar Station, Svalbard. Nine crew members three of whom were women, took part in the study. They filled in the Profile of Mood States (POMS) and the Symptom Checklist 90-Revised (SCL-90-R) after their arrival, at equinox, during polar night, in spring and during the midnight sun. Depression and hostility were highest in the spring following the isolation of polar night. Vigor reached its lowest point in spring and remained low until mission completion. Confusion continued to decline throughout the mission. The Polish crew adapted by monitoring their feelings to work together and ensure survival. Up until and during the polar night, negative feelings were low. Following the isolation period, depression and hostility increased while vigor declined. This suggests adaptation paradigm wherein the participants stopped to monitor their own feelings as closely after the polar night.

21.1 Introduction

Research in the polar social sciences has often taken one of these two approaches: research on isolated and confined environments (ICE) at Antarctic stations, and research on Arctic indigenous peoples. The doctoral research outlined in this article

A.G.M. Temp (✉)
The University of Edinburgh, Edinburgh, Scotland, UK

Institute of Geophysics, The Polish Academy of Sciences, Warsaw, Poland
e-mail: atemp@ed.ac.uk

B. Lee • T.H. Bak
The University of Edinburgh, Edinburgh, Scotland, UK

© The Author(s) 2017
K. Latola, H. Savela (eds.), *The Interconnected Arctic — UArctic Congress 2016*,
Springer Polar Sciences, DOI 10.1007/978-3-319-57532-2_21

addresses a gap in the existing research: research on ICE at Arctic research stations. Stressors that have been identified in the Antarctic environment such as dangers from crevasses, blizzards and the continuous darkness in austral winter (Palinkas and Suedfeld 2008) apply also to the Arctic, even though Arctic blizzards are fewer and weaker and the average temperature is higher than in the Antarctic (Palinkas 1991, 1990). An additional stressor are polar bears which pose a significant threat to human life (Norwegian Polar Institute 2005); they are a uniquely Arctic stressor that is entirely absent in Antarctica. Previous research has shown that Antarctica's social environment has predominant influence on mental health and well-being (Bhatia and Pal 2012). The key stressors at research stations include lack of privacy, boredom, sexual and emotional deprivation, contrived opportunities for social interaction, and reduced possibility to escape or avoid interpersonal conflict and resultant stressful situations (Palinkas 1990).

The Arctic ICE research described in this article aims to gather specific psychological knowledge relevant to mental health and well-being in Arctic personnel. Such knowledge may assist in the selection of Arctic personnel and may also indicate directions for providers of psychological support and self-care of this personnel *in situ.* Discovering the nature of psychological fluctuations and their chronological characteristics in the Arctic circle opens up the possibility of delineating the person- and situation-specific indices of mental health and well-being along with the relevant support.

21.2 Methodology

In this article, preliminary data is reported from nine of the 11 winter team members (comprising three women) at the Polish Polar Station, Hornsund, Svalbard. The winter team arrives at the station by early July each year and remains until late June the following year. The members are isolated at the station from late November to early March each year, with the final sunset before polar night occurring in late October. The presented participants were at the station from early July 2015 to late June 2016. Only one, the station commander, had wintered at the station before but three others had had shorter visits to the station. However, two winter team members are not accounted for in the present report: one withdrew from the study, and one had to be evacuated due to psychiatric complications. Each team member's mood and mental health was assessed using the Profile of Mood States 2 – Brief Version (POMS) and Symptom Checklist 90 Revised (SCL-90-R) at five "Mission Time" points: July ("After Arrival"), September ("Autumn"), January ("Winter"), April ("Spring") and June ("Summer"). The POMS assesses the following subscales: Anger-Hostility, Depression-Dejection, Fatigue-Inertia, Tension-Anxiety, Confusion-Bewilderment and Vigor-Activity, while the SCL-90-R measures the following symptoms of psychopathology: Somatization, Interpersonal Sensitivity,

Obsessive-Compulsive Behaviour, Anxiety, Phobic Anxiety, Paranoid Ideation, Psychoticism, Depression and Hostility.

ICE research faces some issues with data analysis due to the very small sample sizes of the studies. Frequently, studies report a sample size less than 20 (e.g. Palinkas et al. 2001; Reed et al. 2001; Xu et al. 2003) or less than 10 (e.g. Corbett et al. 2012; Leon and Scheib 2007). Studies with larger sample sizes often report longitudinal data spanning several years to several decades (e.g. Bhatia et al. 2013; Palinkas et al. 2000).

Statistical analysis of small samples is potentially problematic as a single outlier may disproportionately affect the findings. As this study reports on a sample of nine participants, details on the analytic procedure are included for transparency reasons. The independent variable (IV) was Mission Time, which was used to predict the dependent variables (DV). The DV consisted of the POMS and SCL-90-R subscales listed above. First, the normality of all DV was assessed using the Shapiro-Wilk test. In case of non-normally distributed DV, the non-parametric *Friedman test* was chosen, followed by group comparisons using *Wilcoxon signed rank tests* with the effect size *r* (Field 2009, p. 579–580). For normally distributed DV, a within-subjects *analysis of variance (ANOVA)* with a *Huyn-Feldt correction* was chosen because Stiger et al. (1998) have shown that this is robust under conditions of a small sample size and ordinal data. For ANOVA effect size, *omega squared (ω^2)* will be reported because it is reliable with small sample sizes (Levine and Hullett 2002). This approach allows two types of conclusions: firstly, to reject that there is no effect of Mission Time on Mood or Mental Health at all (i.e. reject the so-called "null-hypothesis"), and secondly to explain the variance in Mood or Mental Health by Mission Time. In order to accept the so-called "alternative hypothesis" that Mission Time affects Mood and Mental Health, *Bayes Factors (BF)* will be employed, which quantify how much more likely the presented data is to occur under this alternative hypothesis.

21.3   Results

This article focuses on the results on Hostility and Depression indicators from the SCL-90-R, as well as on the Confusion and Vigor indicators from the POMS. Depression refers to dysphoric mood, signs of withdrawal and loss of vital energy, while Hostility describes aggression, irritability and rage. Confusion includes a lack of cognitive or behavioural clarity and a disruption of awareness; Vigor means high levels of psychological and physical energy.

Depression had three levels which were non-normally distributed (Autumn ($W_{(9)}$ = .794, p = .018), Winter ($W_{(9)}$ = .735, p = .004), Summer ($W_{(9)}$ = .631, p = .000); Hostility had four non-normally distributed levels (After Arrival ($W_{(9)}$ = .390, p = .000), Autumn ($W_{(9)}$ = .831, p = .046), Winter ($W_{(9)}$ = .723, p = .003), Spring

$(W_{(9)} = .801, p = .021)$). Such non-normal distributions indicate that the data set contains very heterogenous individual scores so that an ANOVA may not be used. Subsequently, the non-normally distributed Depression and Hostility DV were analysed using Friedman's test and the normally distributed Confusion and Vigor DV with Huyn-Feldt-corrected ANOVA.

Mission Time had a significant main effect on Hostility ($\chi^2_{(4)} = 12.79$, p = .012). Participants felt more hostile in Autumn (M = 1.22, SD = 1.48, p = .039, r = −.307), Spring (M = 2.22, SD = 2.49, p = .033, r = −.625) and in Summer (M = 2.11, SD = 2.26, p = .016, r = −.948) compared to After Arrival (M = 0.22, SD = 0.67). See Fig. 21.1 below.

However, when a Bonferroni correction for multiple comparisons was applied to these Wilcoxon signed rank tests (p = .0125), none of them remained significant. The null hypothesis is always that there is no effect of Mission Time on any given mood or mental health symptom, so if Mission Time had no effect on Hostility, the result above would only have a 1.2% chance of occurring. Mission Time further explains 30.7%, 62.5% and 94.8% of variability in Hostility in Autumn, Spring and Summer, respectively. The Bayesian analysis provided moderate support on the effect of Mission Time on Hostility (BF = 3.34). The combination of these analyses allows to accept that there is a moderate influence of Mission Time on Hostility which would be unlikely to occur if Mission Time did not affect Hostility at all.

Mission Time had a significant effect on the measure of Depression ($\chi^2_{(4)} = 12.76$, p = .012). Participants reported feeling more depressed in Autumn (M = 4.89, SD = 4.56, p = .043, r = −.301) and in Spring (M = 7.33, SD = 6.40, p = .020, r = −.648) than After Arrival (M = 2.33, SD = 1.87), see Fig. 21.2.

After a Bonferroni correction (p = .0125) for multiple comparisons, none of the Wilcoxon signed rank tests remained significant. This shows an effect of Mission Time on Depression in this data that would only have a 1.2% chance of occurring if Mission Time had no influence on Depression at all. Mission Time further explained 30.1% of variability in Depression in Autumn and 64.8% in Spring. The Bayesian analysis provided moderate support of the effect of Mission Time on Depression (BF = 3.84). The combination of these analyses allows to reject the notion that Mission Time does not affect Depression and accept that its influence is moderate.

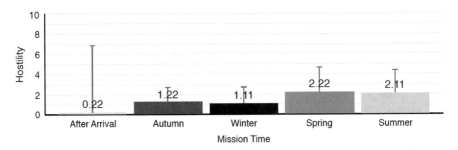

Fig. 21.1 The SD showed greatest variation after arrival, while hostility was highest in Spring and Summer. Nevertheless, the maximum possible hostility was 24, which means participants never reported more than 9.25%

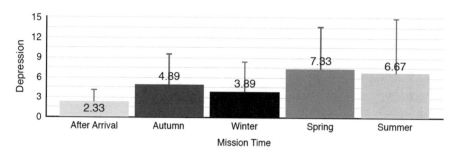

Fig. 21.2 The SD of depression showed greatest variation in Summer. The maximum possible depression was 24, which means participants never reported more than 30.54%

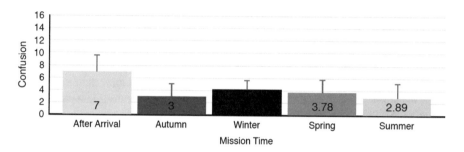

Fig. 21.3 Confusion continued to decline over time, the highest level was 43.75% at the after arrival measuring point. Its SD remained comparatively stable

Mission Time showed a significant main effect on Confusion ($F_{(2.44, 19.51)} = 3.66$, $p = .001$, $\omega^2 = .451$). Participants reported less Confusion in Autumn (M = 3.0, SD = 2.12, $p < .001$), Winter (M = 4.22, SD = 1.56, $p = .016$), Spring (M = 3.78, SD = 2.11, $p = .003$) and Summer (M = 2.89, SD = 2.32, $p < .001$) compared to After Arrival (M = 7.0, SD = 2.65), as indicated by Bonferroni post-hoc tests (Fig. 21.3).

The within-subjects ANOVA implies that in this data, Mission Time had an effect on Confusion that would only have a 0.01% chance of occurring if there were no such effect at all. The Bayesian analysis provided extreme evidence (Andraszewicz et al. 2015) in support of the effect of Mission Time on Confusion (BF = 476.36). The effect of Mission Time on Confusion was thus accepted. Mission Time further explains 45.1% of variability in Confusion.

Mission Time had a significant main effect on Vigor ($F_{(4, 32)} = 2.75$, $p = .045$, $\omega^2 = .159$). Bonferroni post-hoc tests did not show any specific monthly differences (Fig. 21.4).

The within-subjects ANOVA implies that in this data, Mission Time had an effect on Vigor that would only occur with a chance of 4.5% if overall, Mission Time did not influence Vigor at all. The Bayesian analysis provided anecdotal support of the effect of Mission Time on Vigor (BF = 1.72). The combination of these analyses allowed to accept that there is a mild influence of Mission Time on Vigor. Mission Time further explains 15.9% of variability in Vigor.

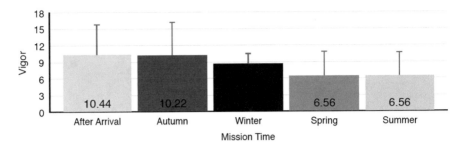

Fig. 21.4 Vigor continuously declined. We have set the y-axis to 18 to accommodate for the SD of after arrival and Autumn but the actual maximum is 16. So the highest Vigor reported was 65.25%

21.4 Discussion

The results show that time spent at the Polish Polar Station comes with a mild decline in vigor, with participants feeling most vigorous right after their arrival and least vigorous just before their departure. A stronger, moderate influence of time spent at the Station on hostility and depression. Both of these were lowest after arrival and peaked in spring, after the end of the isolation. The mental state variable most susceptible to time spent at the Station was confusion: confusion was highest right after arrival and continued to decline as the mission went on. It spiked briefly in winter when participants had to adjust to polar night but proceeded to decline.

These findings imply that adaptation patterns include a decline in both liveliness and confusion: as people spend time at the station they become less excited about and more habituated to their environment. With the environmental change of polar night, confusion increases once more before declining until the end of the mission. During the winter interviews, the participants described great difficulties with monitoring their own feelings as a means to keeping the peace among the team. These interviews help with interpreting the data on hostility and depression: while adaptation continues, people become less able or less willing to monitor their feelings of hostility and unhappiness after the isolation. The interview data suggest that some participants considered that the most difficult part of the mission was behind them after the polar night; hence the observed pattern in hostility and depression with declining monitoring of feelings and more openly hostile behaviour.

The results provide an interesting contrast with research from the Antarctic. Frequently, the highest levels of depression, hostility and anxiety at Antarctic stations are reported during mid-winter (Palinkas and Suedfeld 2008). However, these studies include American, British, Japanese and Chinese nationals, in contrast with the present findings based on Polish nationals. Palinkas et al. (2004) report no POMS fluctuations over mission time for the Polish *Henryk Arctowski Station* on King George Island, Antarctica (62.16° S, 58.47° W). Their findings contrast with the ones here that observed slight fluctuations in confusion and vigor using the same measure. However, it is noteworthy that Arctowski is sub-Antarctic and does not experience polar night during its winter, possibly explaining the different results.

However, all crews of all nationalities have thus far been affected but to different degrees. Individual characteristics that make individual people more prone to winter depression include summer depression and being married (Palinkas et al. 1995).

The fore mentioned issues underline both the strengths and weaknesses of this study. To our knowledge, this is the first study focusing on Polish nationals at an isolated research station during polar night. As such it extends the knowledge of polar psychology, though requires some caution in generalising beyond the sample size used here. Especially so because the effect sizes varied greatly: the lowest variability explained by mission time was 15.9% for overall vigor, the highest was 94.8% for summer hostility. Simultaneously, the low BF for vigor only supplied anecdotal evidence (Andraszewicz et al. 2015) for the influence of mission time. This suggests that one or more other factors than mission time contribute to vigor. Wood et al. (2000) described numerous positive, salient experiences that could explain more of the variability in vigor. These experiences include a rewarding work life as well as field trips. For all of the unpleasant mental states, the variability explained was above 30%, implying that mission time influences explains mental states better than it does positive ones. This means that ICE missions time by itself may have a more negative than positive influence on mental health. Positive experiences during these missions are related to other factors than the mission progressing.

It is concluded that Polish ICE crews in the Arctic experience different adaptation patterns and mental health fluctuations to crews stationed in the Antarctic, regardless of their nationalities. While this evidence may have limited potential for generalization, it is important to collect evidence from many different ICE and nationalities. Ultimately, this kind of knowledge can inform us about selection techniques, support strategies and coping preparations for when mankind begins long-duration spaceflights. It also contributes to improving the understanding on current ICE missions. Future research needs to focus on in-mission emotional and social support and coping strategies to make missions safer and more successful. Knowledge about different nationalities' behavioural variations under ICE conditions is valuable when assembling international crews, because some cultures may be better suited for shared missions.

References

Andraszewicz S, Scheibehenne B, Rieskamp J, Grasman R, Verhagen J, Wagenmakers E-J (2015) An introduction to Bayesian hypothesis testing for management research. J Manag 41:521–543. doi:10.1177/0149206314560412

Bhatia A, Pal R (2012) Morbidity pattern of the 27th Indian Scientific Expedition to Antarctica. Wilderness Environ Med 23:231–238.e2. doi:10.1016/j.wem.2012.04.003

Bhatia A, Malhotra P, Agarwal AK (2013) Reasons for medical consultation among members of the Indian scientific expeditions to Antarctica. Int J Circumpolar Health 72. doi:10.3402/ijch.v72i0.20175

Corbett RW, Middleton B, Arendt J (2012) An hour of bright white light in the early morning improves performance and advances sleep and circadian phase during the Antarctic winter. Neurosci Lett 525:146–151. doi:10.1016/j.neulet.2012.06.046

Field A (2009) Discovering statistics using SPSS, 3rd edn. SAGE Publications Ltd, Los Angeles
Leon GR, Scheib A (2007) Personality influences on a two-man Arctic expedition, impact on spouse, and the return home. Aviat Space Environ Med 78:526–529
Levine TR, Hullett CR (2002) Eta squared, partial eta squared, and misreporting of effect size in communication research. Hum Commun Res 28:612–625. doi:10.1111/j.1468-2958.2002. tb00828.x
Norwegian Polar Institute (2005) Polar Bears in Svalbard
Palinkas LA (1990) Psychosocial effects of adjustment in Antarctica – lessons for long-duration spaceflight. J Spacecr Rocket 27:471–477. doi:10.2514/3.26167
Palinkas LA (1991) Effects of physical and social environments on the health and well-being of Antarctic winter-over personnel. Environ Behav 23:782–799. doi:10.1177/0013916591236008
Palinkas LA, Suedfeld P (2008) Psychological effects of polar expeditions. Lancet 371:153–163. doi:10.1016/S0140-6736(07)61056-3
Palinkas LA, Cravalho M, Browner D (1995) Seasonal variation of depressive symptoms in Antarctica. Acta Psychiatr Scand 91:423–429. doi:10.1111/j.1600-0447.1995.tb09803.x
Palinkas LA, Gunderson E, Holland AW, Miller C, Johnson JC (2000) Predictors of behavior and performance in extreme environments: the Antarctic space analogue program. Aviat Space Environ Med 71:619–625
Palinkas LA, Reed HL, Reedy KR, Van Do N, Case HS, Finney NS (2001) Circannual pattern of hypothalamic–pituitary–thyroid (HPT) function and mood during extended antarctic residence. Psychoneuroendocrinology 26:421–431. doi:10.1016/S0306-4530(00)00064-0
Palinkas LA, Johnson JC, Boster JS, Rakusa-Suszczewski S, Klopov VP, Fu XQ, Sachdeva U (2004) Cross-cultural differences in psychosocial adaptation to isolated and confined environments. Aviat Space Environ Med 75:973–980
Reed HL, Reedy KR, Palinkas LA, Van Do N, Finney NS, Case HS, LeMar HJ, Wright J, Thomas J (2001) Impairment in cognitive and exercise performance during prolonged Antarctic residence: effect of thyroxine supplementation in the polar triiodothyronine syndrome. J Clin Endocrinol Metab 86:110–116. doi:10.1210/jcem.86.1.7092
Stiger TR, Kosinski AS, Barnhart HX, Kleinbaum DG (1998) Anova for repeated ordinal data with small sample size? A comparison of anova, manova, wls and gee methods by simulation. Commun Stat Simul Comput 27:357–375. doi:10.1080/03610919808813485
Wood J, Hysong SJ, Lugg DJ, Harm DL (2000) Is it really so bad? A comparison of positive and negative experiences in Antarctic winter stations. Environ Behav 32:84–110. doi:10.1177/00139160021972441
Xu C, Zhu G, Xue Q, Zhang S, Du G, Xi Y, Palinkas LA (2003) Effect of the Antarctic environment on hormone levels and mood of Chinese expeditioners. Int J Circumpolar Health 62(3):255–267

Part IV
Arctic Tourism

Chapter 22
Tourism Futures in the Arctic

Patrick T. Maher

Abstract The Arctic is changing; it is ever changing in many social, cultural, economic and environmental ways. This chapter will look specifically at tourism in the Arctic: how has it changed? And how might it change in the future? Since the International Polar Year (IPY) in 2007–2008 there has been a rise of interest in tourism from academia, industry and local communities. Many authors have provided a look into the "deep" past of tourism development; with a number of books and article coming out around 2010, and some have offered thoughts on the future. This chapter will gaze further into the future, to the year 2030. What might the growth in tourism look like, based on the trajectory since 2008 or 2010? What will be the priorities for tourism growth or tourism research in the region?

22.1   Introduction

Tourism is growing and changing; this can be seen in emerging destinations as well as in those, which are now on everyone's "bucket list" of must-sees. Globally more people are travelling, and to destinations that were once far off the typical route. There is a larger world population with the disposable income to travel, and there is a need to forecast some of this growth. Tourism futures, as a type of holistic forecasting, is relatively new. Yeoman's (2012) work gives an excellent overview of this field, and with specific reference to Arctic tourism there are examples of "what if?" and "where are we headed?" – the speculative questions, which have already been examined in relation to some Arctic/Polar nations (see Enger et al. 2015; and others in the *Journal of Tourism Futures*). Tourism in the Arctic is an example of a forecasted change or future by its very nature. The Arctic had been an emerging or unknown destination for many years, but with the scientific focus of the International Polar Year (IPY) and media attention to climate change, more so than ever before

P.T. Maher (✉)
Cape Breton University, Sydney, NS, Canada
e-mail: pat_maher@cbu.ca

© The Author(s) 2017
K. Latola, H. Savela (eds.), *The Interconnected Arctic — UArctic Congress 2016*,
Springer Polar Sciences, DOI 10.1007/978-3-319-57532-2_22

people now want to "see it before it's gone", to use a catch phrase for this type of last chance tourism. We could go back to the beginning of tourism development in the Arctic to start the discussion, but perhaps a more useful starting point lies in the past 10 years.

Ten years ago the world was focused on the Polar Regions through science and the media – this was the International Polar Year (IPY 2007–2008). At that time there was very little attention paid to tourism by the large IPY projects and by various international scientific committees (Maher 2007). Academics did their work in traditional disciplines, and the tourism industry went about business as usual. Tourism was certainly present and had been so for hundreds of years in the Nordic countries. The IPY did, however, give us a tidy starting point from which to examine the future. Maher (2013) provides a thorough run down on the more distant past through the keynote of the 2nd conference of the International Polar Tourism Research Network, so between those two publications (Maher 2007, 2013) it is possible to create a starting point. In 2007 (pp. 3–4), Maher listed some key questions as part of the discussion at the 2006 Canadian Association of Geographer's meeting. A few of these were as follows:

General research issues

- In terms of research, there are many opportunities, but what should be the priorities?
- While the Polar Regions may be 'high profile' again with renewed interest and political will, will it all dry up at some point?
- Is the sense that tourism growth is inevitable correct?
- What are the implications of climate change?

Communication

- Amongst both researchers and operators, what is the scope or rather need for networking opportunities?
- Where can they occur?
- When and under what auspices?

Niche sectors

- Specific to cruise ships, what are the dangers (bigger ships, bigger infra- structure, bigger cultural changes, leakage of $$)?
- What is the social impact of cruise ships, and what is the role of participatory research?

Maher (2013) posited that within each of the three realms (academia, industry and community) there were critical issues to work on. For academia it was celebrating new ideas, creating collaboration, updating publications, and simply cooperating. For industry there was the need to mend bridges, the mistakes of past researchers, and a carefully cultivated future together; one which recognizes good science/social science, but also the practicalities of running a business. For communities it was all about engagement and respectful relationships. So how have some of these questions and concerns been addressed in the subsequent years? This chapter will next attempt to address that, and then forecast for the next 15 years and beyond.

Table 22.1 Estimates of tourist numbers to a variety of Arctic Regions then, as modified from Maher (2013, pp.23–24)

Country/region/province	Tourist numbers (Estimates)	Sources/notes
USA (Alaska)	1,631,500	Summer 2006 data for all out-of-state visitors
Canada		
Yukon	8049	2004 data – covers only the Northern Yukon tourism region
Northwest territories	62,045	2006–2007 data for all non-resident travellers to the entire territory
Nunavut	9,323	2006, summer only
Nunavik (Northern Quebec)	25,000	Nord du Quebec statistics included both the Nunavik and James Bay regions
Nunatsiavut (Northern Labrador)	565	2008 visitors to Torngat Mountains National Park
Greenland	33,000 (air arrivals)	Data reported in 2011
	22,051 (cruise arrivals)	
Iceland	277,800	Data reported in 2008
Svalbard (Norway)	29,813	AECO personal communication, August 2010; 2009 cruise visitors arriving from overseas
Norway (Finnmark)	2,420,959	Data from 2002
Sweden (Norrbotten county)	1,700,000	Data from 2001 tourist overnight stays
Finland (Finnish Lapland)	2,117, 000	2006 data for the number of registered tourist overnights
Russia	Estimated at a few tens of thousands and growing steadily	Actual data difficult to obtain

22.2 Tourism Growth

The growth of tourism in the circumpolar North is perhaps the easiest future metric to map. Is this growth inevitable, as was asked in 2007? Perhaps. Table 22.1 (modified from Maher 2013) showcases some tourism numbers from the mid 2000s, generally from 2006 until 2009 (the IPY years), but in some cases earlier.

Table 22.2 is an updated version covering the same regions, as best as possible, and offering updated estimates.

Using the jurisdictions where there is some consistency in data source and metrics recorded (Alaska, Northwest Territories, Nunavut, Iceland, Sweden and Finland) the future of tourism growth does appear to have been inevitable. Each of these jurisdictions continued on an upward trajectory. The most staggering increase is Iceland with almost a six-fold increase. All tourism markets appear to be growing,

Table 22.2 Estimates of tourist numbers to a variety of Arctic Regions now by using the most current numbers available

Country/region/ province	Tourist numbers (Estimates)	Sources/notes
USA (Alaska)	2,066,800	https://www.commerce.alaska.gov (Accessed January 2017); April 2016 update on 2014-2015 data for all out-of-state visitors
Canada		
Yukon	255,000	http://www.tc.gov.yk.ca (Accessed January 2017); 2015 estimated total overnight visits to the entire territory
Northwest territories	93,910	http://www.iti.gov.nt.ca (Accessed January 2017); 2015-2016 total visitors to the entire territory
Nunavut	16,750	http://nunavuttourism.com (Accessed January 2017); 2015 exit strategy – non-resident visitors
Nunavik (Northern Quebec)	1,000	http://www.tourisme.gouv.qc.ca (Accessed January 2017); 2010 report for 2008 visitor volume in Provincial zone 21
Nunatsiavut (Northern Labrador)	19,840	http://www.btcrd.gov.nl.ca (Accessed January 2017); 2015 accommodation occupancy for Provincial zone 1 (Rigolet-Nain, Labrador)
Greenland	80,862	http://www.tourismstat.gl (Accessed January 2017); 2015 Greenland Tourism statistics for international air departures and number of paid overnights
	218,539	
Iceland	1,289,100	http://www.ferdamalastofa.is (Accessed January 2017) 2015 international visitors to Iceland
Svalbard (Norway)	118,614	http://sysselmannen.no (Accessed January 2017); 2014 Svalbard Reiseliv statistics for overnight stays in Longyearbyen
Norway (Nord Norge – northernmost 3 counties)	1,045,538	http://ec.europa.eu (Accessed January 2017); 2016 data for the number of nights spent at tourist accommodation by a non-resident
Sweden (Norrbotten county)	2,152,000	http://www.lansstyrelsen.se/norrbotten (Accessed January 2017); 2014 *Facts about Norrbotten* report; data from 2013 tourist overnights
Finland (Finnish Lapland)	2,523,897	http://visitfinland.com (Accessed January 2017); 2016 data for the number of registered tourist overnights
Russia	500,000	Tzekina (2014)

and what little decline we see, may be best discussed with regards to the issue of how/where/when statistics are collected. The article by de la Barre et al. (2016) notes this as a foremost issue for the future, one that does impede our ability to forecast and strategically plan. There is now data available from Russia, and also many other jurisdictions, whether country or sub-national region, seem to have improved their data collection systems; although there are still remaining issues with being able to tease out regional subsets versus data from large units (e.g., country-wide).

22.3 Tourism Priorities

In the Arctic, discussion of priorities (or a research agenda) for tourism could be seen to begin with the work of Stewart et al. (2005). They set the initial bar and opened up the conversation again, as earlier researchers in the early 1990s had done so already in a previous wave of interest in polar tourism. This became a step towards much of the action, which has addressed Maher's (2007) questions on communication. Since 2008, there have been five conferences of the International Polar Tourism Research Network (IPTRN), and each of these has progressed the conversation by including new participants. Some conferences had more industry, others more community members; all had a slate of new graduate students depending on the location and year, or former graduate students now with early career positions. While there is no concrete "path" that has been detailed for tourism futures in the Arctic, there is much more recognition of the various possibilities. The IPTRN conferences have led to many joint publications and large-scale research projects, but recognizing that tourism is such a broad field of research an agenda cannot possibly contain everything at the same time. Thus, a true joint research agenda may be an impossibility. The future of the IPTRN appears solid with conferences already planned for the Yukon (2018) and Tierra del Fuego (2020). The corresponding University of the Arctic Thematic Network on Northern Tourism has also seen growth and expansion, which should continue. As Maher (2013) notes, the Thematic Network on Northern Tourism was founded to bridge the teaching-research barrier and when started in 2008 had big ideas, but little functional support. It is now one of the largest thematic networks in the University of the Arctic network, with more than 20 partners as of 2017, and its flagship program, a joint masters curriculum, has begun to take shape. A SIU funded pilot allowed seven institutions in five Arctic nations to implement a field course to Eastern Finnmark and two additional online courses in 2016/2017 – bringing together more than 40 faculty and students from 17 countries.

Topics such as climate change have become increasingly important in Arctic tourism, and the academic community has responded with insightful empirical research abounding (see Dawson et al. 2007, 2010; Kajan 2014). Local communities have become more and more engaged – as was deemed critically important by Maher (2013). This has occurred in almost every scientific discipline since the IPY. Examples of community-based citizen science in relation to tourism, which provides true feedback to the community are documented by de la Barre et al. (2016) in locations across the North, but still more could be done.

With regards to niche sectors, cruise tourism has certainly become a critical player. In 2007 when the MS Explorer sank, it opened up many people's eyes to the true dangers possible in Arctic cruise tourism (see Stewart and Draper 2008). In the Canadian Arctic, the concern was that a similar incident could happen near a local community without any possibility of assistance due to aging infrastructure and/or non-existent monitoring. In 2010, when the Clipper Adventurer grounded (see Stewart and Dawson 2011) there were new fears of the same kind, yet the largescale voyage of the Crystal Serenity took hold in 2016. By all accounts the future involving larger cruise ships, seeking passage through key routes is upon us, and for tourism the attractions, such as the discovery of Franklin's ships and subsequent media attention, is upon us as well. More large ships will come, that is certain with Crystal Cruises (owners of the Crystal Serenity) already planning additional transits in summers 2017 and 2018. The real concern now is whether other operators will give as much thought for the environment (having an additional icebreaker accompany the voyage) or culture/society (through extensive pre-trip consultation), and thus due diligence to the undertaking; and when or if the Canadian management/permitting system will catch up to its European counterparts (using AECO and the Governor of Svalbard as an example).

Another niche tourism sector, which is very much growing, is Indigenous tourism. Canada has chosen to focus much of its marketing on Indigenous (Aboriginal) products and attractions, particularly in the Arctic and provincial peripheries. This focus is echoed through new governance regimes (autonomy, self-government, etc.) that give Indigenous communities more say and engagement in many regions of the Arctic. The academic community is also at the crest of this wave with new work such as that by Viken and Müller (2017).

22.4   Conclusions – Futures Towards 2030

The year 2017 has been named the "International Year of Sustainable Tourism for Development" by the UN General Assembly, so what better time to forecast the future than now. Overall, the growth, communication and engagement in Arctic tourism are on positive trajectories. There are some concerns around carrying capacity of vulnerable Arctic areas; for example, can Iceland both culturally and environmentally, sustain such continued growth? There are also concerns around the "slippery slope" entered in regards to large cruise vessels. However, more than ever there is enthusiasm for new ideas and new technologies, and there is starting to be some proof that better networks lead to better results; collaboration is a good thing (see Stewart et al. 2016). Cooperation will manifest through the loss of national or academic politics – creating synergies vs. isolationism and leading to full engagement with a full suite of industry and community partners.

Governance issues and skepticism of industry-academia collaboration should subside and there are already many examples whereby tourism and science can go hand in hand. Industry associations are growing, which is good – especially when the most promising practices are shared. This is most evident in the recent move by the Association of Arctic Expedition Cruise Operators (AECO) into Canada, as well as the interest of the Arctic Council's working group on the Protection of the Arctic Marine Environment (PAME) in more circumpolar cruise guidelines. However, no group such as the Arctic Council or AECO, is forgetting the need for grassroots and sub-jurisdictional buy in – a laddering of planning strategies from circumpolar to national and to regional levels. The development towards 2030 will be interesting.

Hopefully, the International Year of Sustainable Tourism for Development can be a starting point for another five IPTRN conferences, where more people safely and securely see the region, and where all facets of the tourism industry develop. There will as well be a new suite of educators and researchers if some of the University of the Arctic activities prevail until 2030.

The comparative work done by Maher et al. (2014) has revealed many possible directions for future research, which could move Arctic tourism in a more sustainable direction – corresponding to the UN designated year. The strain between the perceived need for economic development through tourism (and the resultant demand for more infrastructure) and the fear that more tourism will degrade natural environments and negatively impact small communities will continue. There is no automatic or standardized solution to this, and every location will feel the strain differently. As research offers better proof – for example in a comparison of governance of tourism in multiple Arctic countries, jurisdictions will not be able to fully understand all the possible public, private, and civic stakeholder roles in the development of tourism.

Acknowledgements This chapter would not have been possible without many ongoing conversations and thoughtful discussions amongst colleagues, found in both the University of the Arctic Thematic Network on Northern Tourism and International Polar Tourism Research Network. These networks, founded at the end of the IPY, showcase the "good" in university-industry-community tripartite work. They are not perfect, but certainly deserve more recognition than they currently receive.

References

Dawson J, Maher PT, Slocombe DS (2007) Climate change, marine tourism and sustainability in the Canadian arctic: contributions from systems and complexity approaches. Tour Mar Environ 4(2–3):69–83

Dawson J, Stewart E, Lemelin H, Scott D (2010) The carbon cost of polar bear viewing in Churchill, Canada. J Sustain Tour 18(3):319–336

de la Barre S, Maher PT, Dawson J, Hillmer-Pegram K, Huijbens E, Lamers M, Liggett D, Müller D, Pashkevich A, Stewart EJ (2016) Tourism and arctic observation systems: exploring the relationships. Polar Res 35. http://dx.doi.org/10.3402/polar.v35.24980

Enger A, Sandvik K, Iversen EK (2015) Developing scenarios for the Norwegian travel industry 2025. J Tour Futures 1(1):6–18

Kajan E (2014) Arctic tourism and sustainable adaptation: community perspectives to vulnerability and climate change. Scand J Hosp Tour 14(1):60–79

Maher PT (2007) Arctic tourism: a complex system of visitors, communities, and environments. Editorial foreword to special issue of Polar Geogr, 30(1–2), 1–5

Maher PT (2013) Looking back, venturing forward: Challenges for academia, community and industry in polar tourism research. In: Müller DK, Lundmark L, Lemelin RH (eds) New issues in polar tourism: communities, environments, politics. Springer, Amsterdam, pp 19–36

Maher PT, Gelter H, Hillmer-Pegram K, Hovgaard G, Hull J, Jóhannesson GT, Karlsdóttir A, Rantala O, & Pashkevich A (2014) Arctic tourism: realities and possibilities. 2014 Arctic Yearbook, 290–306

Stewart EJ, Draper D, Johnston ME (2005) A review of tourism research in the polar regions. Arctic 58(4):383–394

Stewart EJ, Dawson J (2011) A matter of good fortune? The grounding of the *Clipper Adventurer* in the northwest passage, Arctic Canada. Info North, Arctic 64(2):263–267

Stewart EJ, Draper D (2008) The Sinking of the MS Explorer: implications for Cruise Tourism in Arctic Canada. Info North, Arctic 61(2):224–228

Stewart EJ, Liggett D, & Dawson J. (2016) The maturing of polar tourism as a 'field of study'. Presentation at the 5th conference of the International Polar Tourism Research Newwork, Akureyri, Iceland

Tzekina M (2014) Estimation of tourism potential of Russian Far North. PhD, Economic, social, political and recreational geography. Moscow State University: Moscow.

Viken A, Müller D (eds) (2017) Tourism and indigeneity in the Arctic. Channel View, Bristol

Yeoman I (2012) 2050: Tomorrow's tourism. Channel View, Bristol

Chapter 23
Uniqueness as a Draw for Riding Under the Midnight Sun

Blake Rowsell and Patrick T. Maher

Abstract The Yukon Territory, in Canada's western Arctic, has tremendous potential for tourism. The territory has abundant natural beauty, and a historical mystique that naturally draws people to the destination. The Yukon has already established itself as an adventure tourism destination for activities such as paddling (canoeing, kayaking and rafting) and dog-sledding. However, mountain biking is a new segment of the Yukon tourism industry. A greater understanding of tourism in the Arctic can be developed through an examination of the destination attributes that draw mountain bike tourists to the Yukon.

23.1 Introduction

Mountain biking is one of the most popular recreational activities worldwide (Leberman and Mason 2000; Taylor 2010). Research in the United States has shown that annually cycling contributes $133 billion to the U.S. economy, supports nearly 1.1 million jobs, and provides sustainable growth in many rural communities (Newsome and Davies 2009; Outdoor Industry Foundation 2006).

Destination attributes are the on-the-ground tangible features of a destination; the site characteristics that are critical in a tourists' vacation destination decision-making (Weaver 1994). For mountain biking these attributes include inherent features such as natural landscapes, as well as constructed amenities such as trails (Freeman 2011). Destination attributes are the pull factors that make up the attractiveness of a region (Chon 1990; Devesa et al. 2010), and the factors by which a person is motivated to want to visit (Fluker and Turner 2000) the site. The Yukon is renowned for its beautiful scenery, and the lack of development providing an

B. Rowsell (✉)
Capilano University, North Vancouver, BC, Canada
e-mail: blakerowsell@capilanou.ca

P.T. Maher
Cape Breton University, Sydney, NS, Canada
e-mail: pat_maher@cbu.ca

© The Author(s) 2017
K. Latola, H. Savela (eds.), *The Interconnected Arctic — UArctic Congress 2016*,
Springer Polar Sciences, DOI 10.1007/978-3-319-57532-2_23

"excess" of nature. Arctic destinations, such as the Yukon, can capitalize on tourism dollars by taking advantage of these natural amenities.

23.2 Methodology

This project was guided by initial meetings with key informants in the Yukon mountain bike community. These contacts included the owners of Boréale Adventures, the head of the Whitehorse municipal trail crew, individual members of the trail crew, the executive of the local mountain bike club, and other local mountain bikers. In addition to being guided by stakeholder involvement, this research project also concluded by conducting feedback sessions and formal meetings to assist with closure of the project, and to ensure the return of information and dissemination of findings to the community. This practice follows the work of Kindon et al. (2007).

Qualitative research methods were used to understand a larger reality in the context of the trends and patterns occurring in mountain bike tourism in the Yukon. The project used semi-structured interviews, and participatory observation. Interviews were conducted on site at Boréale Adventures with the clients of the mountain bike–themed eco-lodge and with other mountain bike tourists in Whitehorse. Participant observation in the form of "ride alongs" were also conducted with mountain bike tourists in Whitehorse. These dual methods allowed for greater insight into motivations and interests that people may be unwilling to talk about or that would have been missed otherwise (Taylor 2010). In addition, it allowed for an understanding of the context (Patton 2002) for this research.

The participants were all on a mountain bike vacation in Whitehorse. Many were clients of Boréale Adventures, although some participants were also on independent holidays. Participants ranged in age from 25 to 62, and 71% of the study participants were male. Participants ranged from "dirtbag" free and independent travellers who were camping or sleeping in their car to make the trip as cheap as possible, to those that had flown thousands of kilometres and desired amenities like steak and champagne on the trail. All participants were asked about their motivations and feelings with respect to mountain bike tourism. All participants interviewed had been involved in mountain bike tourism before their trip to the Yukon.

23.3 Tourist Preferences

Remoteness and scenic attributes of the destination are key draws for mountain bike tourists. However, the tourists traveling to the Yukon Territory were in search of something specific: they desired the uniqueness of the destination.

Mountain bike tourists felt that there had to be characteristics of a destination that were different than they were used to. Participants did not want their mountain bike vacation to be like riding at home; most riders indicated that they have enjoyable trails at home, so when they travel they want to experience something more out of the destination. One mountain bike tourist explained that

> if you travelled miles and miles, and you rode a trail that was as good as one of your local trails but felt like one of your local trails, it would be a disappointing trail. Because, why did I travel all that distance?

Mountain bike tourists seek experiences that provide lasting memories (Young 2008). In particular, mountain bike tourists sought something different than what they were used to through a trip to the circumpolar north. Mountain bike tourists were searching for something to write home about. They emphasized that when a destination was unique, they were excited to tell people how great the experience was. Travel to the Arctic is not an easy task, and as a result tourists invest a lot of time and money into their trip. Mountain bike tourists expressed a desire for an experience that is different from what they normally have. The uniqueness can come from a variety of different factors or attributes, but something has to be unique.

Participants attributed many factors needed to create this uniqueness for them. It could be something as simple as different trees, or a different trail surface than they are used to riding. Many participants indicated that different scenery could also differentiate one experience from another.

In many regions around the world, signature bike trails have come to define a region's mountain bike culture (Young 2008). Signature trails help to strengthen a regions' trail network and draw tourists into a destination. Typically, signature trails offer unique backcountry experiences, with stunning scenery and offering a sense of physical accomplishment (Young 2008). Many of these trails have been granted "epic" status by the International Mountain Bike Association (IMBA). IMBA claims that "if you are a mountain biker, this is your bucket list. Every single one of the mountain bike trails listed on this page will blow your mind. Guaranteed" (IMBA 2012). Many study participants indicated a clear link between IMBA's bucket list and their personal to-do lists. Mountain Hero trail in Carcross, just outside Whitehorse was granted epic status in 2011; IMBA states that the Mountain Hero trail offers stunning alpine views, historic mining artefacts, and a chance to see caribou and other wildlife (IMBA 2012). Figure 23.1 is an example of the mining history that is located along the Mountain Hero trail. Many study participants indicated that riding the Mountain Hero trail was the highlight of their trip. The majority of the participants had heard about the Mountain Hero trail before coming to the Yukon; however, it exceeded everyone's expectations.

The concept of a "bucket list," a list of to-do things that riders need to cross off, came up as a motivation for travel to the Yukon. Several riders desire to do and see things that they have not experienced yet in their lifetime. Riders suggested that "I don't want to do something I've done before. It's gotta be new." This characteristic of uniqueness was seen as being a defining factor, or at least a value-added feature, of a destination.

All participants highlighted the importance of scenic appeal as a unique destination attribute. Some riders even went as far as suggesting that scenery is one of the primary motivating factors for participation in the sport. The term scenery had

Fig. 23.1 Mining history on Mountain Hero Trail (Photo courtesy: B. Rowsell)

different meanings for different participants. However, they all indicated that the overall concept of scenery is very important to rider satisfaction about a tourism destination. One mountain bike tourist explained that "scenery is important; otherwise, I can just stay in my own backyard to ride." Generally, participants indicated that scenic appeal makes for a great riding experience. They highlighted the importance of scenic views of mountains, and lakes, and suggested that scenic views help them put their life into perspective within the bigger picture. Figure 23.2 is an example of scenery around Carcross, Yukon. The participants also indicated the importance of scenic nature, including forests, trees, and flowers. The concept of scenery discussed by interviewees was divided into two thematic categories: views and nature. Views being when trails take riders out of the forest, and allow them a view of the larger surroundings. Nature being the natural things that riders see while in the forest, or on the trail.

At lower latitudes around the world, the majority of mountain biking takes place on trails surrounded by trees. While participants indicated that they enjoy riding in forest settings, many suggested that alpine views are important to enhancing their riding experience and making it different that they are used to. Participants discussed that scenery is not just trees; they desired to see for miles in front of and

Fig. 23.2 The view from the top of Montana Mountain, Carcross, Yukon (Photo courtesy: B. Rowsell)

behind them while riding. Many of the interviewees explained that they are attracted to mountains. Many Arctic regions allow riders to experience scenic views such as mountains, which riders identified as being spectacular to look at. Also, the increase in elevation allowed by mountains permits access to viewpoints with unobstructed views. This access is made easier by the lower tree line in Arctic regions. Most riders emphasized the appeal of alpine riding above the treeline for both the uniqueness and scenery.

Participants indicated that uniqueness can also come from the remoteness and mystique of a destination. For many riders, remote areas were also seen as pristine. One mountain bike tourist indicated that one of his favourite characteristics about riding in the Yukon was the feeling of remoteness. Figure 23.3 is an example of the remoteness of the Yukon River. Many mountain bike tourists pointed out that this sense of remoteness could be accomplished by both scenic views, or through forest settings. Participants also explained that they desire to see nature as it was meant to be – in its natural state. While on trails, riders in this study were looking for scenery they considered to be pretty and unique, which included forests, trees, nature, and wildflowers. Some tourists mentioned that they appreciate seeing wildlife on trails, and enjoy forests and wooded areas because it increases their chance of seeing wildlife. Participants explained that they enjoy viewing wildlife that they could not view at home, such as caribou and grizzly bears.

Fig. 23.3 Remoteness on the Yukon River Trail (Photo courtesy: A. Campbell)

Mystique can be related to the culture or history of a destination. Participants highlighted, though, that this type of uniqueness does not have to be on or near the trails but part of their overall tourism experience. Mountain bike tourists indicated that a nearby town could provide the desired uniqueness if it was rustic or authentic. In addition to feeling remote and having a different culture, riders said they seek out places that are unrefined, and create what they feel is a true adventure because it feels a little unsafe, giving it an edgy feel.

The quality of a mountain bike–specific holiday can best be determined by the mountain bike rides that participants take while on their trip. In particular, the quality and variety of trails ridden, and the scenic appeal of the trails add to the quality of the holiday as a whole. However, the participants required more in a great trip – they wanted something unique. The uniqueness could be related to the above-mentioned attributes, other factors of the trail, or of the trip itself. Mountain bike trails are simply the vehicle through which mountain bike tourists can experience the remoteness, wilderness, and heritage.

23.4   Implications for Arctic Tourism

The results of this study show that participants were seeking an experience that is different from anything they have done before, but still within their comfort zone. This can be accomplished in the circumpolar north by taking an activity that tourists already know and feel comfortable with, and using it as a vehicle to see or experience something new to the tourist. In the case of the Yukon, trail builders have designed trails based around historic landmarks, or scenic views, or alpine areas, or a sense of remoteness to create a unique tourism experience. Other Arctic destinations may be able to capitalize on existing unique features by designing trails that allow tourists to access attributes that are notable and different. In particular, access to alpine regions, fjords, or historic and cultural areas of significance can create a unique and memorable experience for tourists. To draw tourists in, Arctic destinations can focus image-specific marketing campaigns on demonstrating the uniqueness of a location that can be most readily accessed. As an example, connecting long-distance mountain biking in the summer in Finnmark, Norway to the world-renowned Finnmarksloppet would seem to create the same appeal (Northern Norway Tourism Board n.d.). Combinations of unique features can help create an experience and perhaps play on the participants' sense of adventure.

The Yukon, as a destination, has to offer something unique to draw mountain bike tourists there. The major elements of uniqueness that the Yukon mountain bike tourism industry can capitalize on are wilderness, remoteness, and to a lesser extent, heritage, in order to develop a unique destination for mountain bikers.

Aspiring mountain biking destinations such as Whitehorse need to take advantage of the uniqueness of the destination. Because many riders have such a strong attachment to their home trails, destinations have to work hard to differentiate themselves and make their destination unique. There is no simple formula for uniqueness, as each location has to capitalize on its own characteristics. However, the information from this study indicates that the Yukon already has begun to take advantage of characteristics that are desirable by mountain bike tourists.

23.5 Conclusions

With the growth of the sport as a leisure time activity, the economic impact of mountain bike tourism is beginning to be realized and capitalized on by many destinations. Cycle tourism can be a path towards economic development. Arctic destinations such as the Yukon Territory can capitalize on increased understanding of the importance of destination attributes for mountain bike tourists to increase their tourism market share.

Destination attributes are the on-the-ground tangible features of a destination. Destination attributes help to influence where mountain bikers want to ride (Taylor 2010) and drive the tourist's choice of and substitution between destinations (Moran et al. 2006).

Understanding the attributes that attract mountain bike tourists to a destination can help destinations offer an enjoyable vacation and can assist destinations in capitalizing on the mountain bike tourism market. There are many specific attributes that mountain bike tourists value. Great trails alone are not enough to satisfy mountain bike tourists because there are many great trails worldwide to choose from; destinations have to offer more to give mountain bike tourists something to write home about.

Acknowledgements This chapter has been developed from the MA NRES (Tourism) thesis of the first author (see Rowsell 2013), supervised by the second author. For their assistance throughout the project, we would also like to acknowledge committee members Dr. Philip Mullins and Dr. Steve Taylor. Funding for this research was provided by the Norwegian Ministry of Foreign Affairs, who were interested in destination development across the circumpolar north through the Arctic Chair at Finnmark University College (now the Alta campus of UiT: Arctic University of Norway). In addition, financial support was provided from the Northern Scientific Training Program of the Canadian federal government, and further in-kind support came from the Department of Tourism and Culture – Government of Yukon, and the City of Whitehorse. Lastly, thanks to Marsha and Sylvain at Boréale Adventures, and their clients who shared their holidays.

References

Chon K (1990) The role of destination image in tourism: a review and discussion. Revue De Tour 45(2):2–9

Devesa M, Laguna M, Palacious A (2010) The role of motivation in visitor satisfaction: empirical evidence in rural tourism. Tour Manag 31(4):547–552

Fluker M, Turner L (2000) Needs, motivations, and expectations of a commercial whitewater rafting experience. J Travel Res 38:380–389

Freeman R (2011) Mountain bike tourism and community development in British Columbia: critical success factors for the future (Masters thesis). Royal Roads University

International Mountain Bicycling Association (IMBA) (2012) IMBA epics rides. http://wwwimba-com/epics/rides. Accessed 30 May 2013

Kindon S, Pain R, Kesby M (2007) Participatory action research approaches and methods: connecting people, participation and place. Routledge, London

Leberman S, Mason P (2000) Mountain biking in the Manawatu region: Participants, perceptions, and management dimensions. N Z Geogr 56(1):30–38

Moran D, Tresidder E, McVittie A (2006) Estimating the recreational value of the mountain biking sites in Scotland using count data models. Tour Econ 12(1):123–135

Newsome D, Davies C (2009) A case study in estimating the area of informal trail development and associated impacts caused by mountain bike activity in John Forrest National Park, Western Australia. J Ecotourism 8(3):237–253. doi:10.1080/14724040802538308

Northern Norway Tourist Board (n.d.) Mountain biking over the Finnmark plateau. http://www.nordnorge.com/EN-salten/?News=447

Outdoor Foundation (2006) The active outdoor recreation economy: a $730 billion contribution to the U.S. economy. www.outdoorindustryfoundation.org. Accessed 13 Jan 2017

Patton MQ (2002) Qualitative research and evaluation methods. Sage Publications, Thousand Oaks

Rowsell B (2013) Mountain bike tourism development under the midnight sun: Capitalizing on destination attributes to maximize tourism potential in the Yukon Territory, Canada (MA NRES thesis), University of Northern British Columbia, Prince George, Canada

Taylor S (2010) 'Extending the dream machine:' understanding dedicated participation in mountain biking (PhD thesis), University of Otago, Dunedin, New Zealand

Weaver P (1994) The relationship of destination selection attributes to psychological, behavioural and demographic variables. J Hosp and Leis Mark 2(2):93–109

Young J (2008) Mountain bike tourism: tourism business essentials. http://www.destinationbc. ca/getattachment/Programs/Guides-and-Workshops/Guides/Tourism-Business-Essentials-Guides/MountainBikingTBEGuide2011_May12.pdf.aspx. Accessed 13 Jan 2017

Chapter 24
Arctic Tourism: The Design Approach with Reference to the Russian North

Svetlana Usenyuk and Maria Gostyaeva

Abstract This chapter discusses the potential of design research and education practice to contribute to Polar/Arctic tourism studies. With the geographical reference to the Russian Far North, it is explored what involves in being a human in severe environmental conditions, and what kind of design, clothing, dwelling, transportation it fosters. As a key argument, a perspective is developing of the arctic tourism as an embodied way of (short-term) living in the extreme environment. The discussion is continued by outlining the design approach to Arctic tourism development and based on existing variety of tourism resources in the Russian North two modes of their representation by design – static and dynamic – are suggested. Each mode is further illustrated with a case study of design projects conducted at the Arctic Design School, Yekaterinburg, Russia. To conclude, this chapter offers new ways of understanding and using design as a tool to respond to challenges and opportunities that today's Arctic uncovers not only for tourism, but also for other sectors of Arctic-based and oriented industry.

24.1 Introduction

Today, the tourism industry is one of the most prominent sectors of economy for many countries and communities worldwide. In the Arctic region,[1] despite of its remoteness and strong seasonality as well as the historical dominance of intensive resource exploitation, such as mining, fishing, and oil & gas extraction (Duhaime and Caron 2006; Brigham 2007), the tourism industry shows consistent growth in terms of the quantity of tourists and the portion of income (Maher et al. 2014). However, in Russia, which possesses almost 2/3 of the Arctic territory, the

---

[1] For the purpose of this chapter, the Arctic is defined as in the first Arctic Human Development Report (2004, pp. 17–18), with addition of the middle part of Western Siberia, i.e. Khanty-Mansy autonomous okrug, based on the specific definition of the Russian Far North given by Soviet geographer S.V. Slavin (1972).

S. Usenyuk • M. Gostyaeva (✉)
Ural State University of Architecture and Art, Yekaterinburg, Russia
e-mail: maria_gostyaeva@mail.ru

© The Author(s) 2017   231
K. Latola, H. Savela (eds.), *The Interconnected Arctic — UArctic Congress 2016*,
Springer Polar Sciences, DOI 10.1007/978-3-319-57532-2_24

development of Arctic-based tourism is a new issue in the public strategic planning, research and commercial activities (Kharlampieva 2016).

In this article, the design exploration starts with a statement that the Russian Far North has potential to become a platform for diverse tourism development and, furthermore, to enable new kinds of tourist products and services. Design as a system of material objects and actions aimed at changing existing situations into preferred ones (Simon 1996), can support this trend and provide for embodying new standards of tourism ethics and sustainability in the Arctic region. This article contributes to the studies of Arctic tourism development by changing the perspective on tourism and illuminating new opportunities for transforming given resources into tourist products/services by design.

This study relies on the experience of the Arctic Design School, ADS (Ural State University of Architecture and Art, Yekaterinburg, Russia) obtained during over 30 years of studio and field-based research, teaching and designing for the extreme environment of the Arctic. By focusing on adaptation and sustainability, it is suggested here to consider remote terrains of the Russian Far North as a natural 'living lab' (Bergvall-Kareborn et al. 2009), which provides an exceptional window to what is involved in being a human in severe conditions, and to what innovative design solutions it fosters.

The next paragraph outlines the theoretical foundation for design in Arctic tourism and later on two case studies from collective projects at the ADS are used to illustrate the theoretical background.

24.1.1 Theoretical Foundation for Design in Tourism

Known as one of the greatest producers of experiences (Binkhorst and Dekker 2009), tourism industry constantly calls for efficient 'mechanism' of transforming natural attractions into meaningful memorable experiences of users/tourists. On the other hand, design broadly defined as a field of theory and practice of developing products, services, processes, events, and environments, and particularly its subfield so-called 'experience design' is a unique professional activity with predominant focus on the quality of the human/user experiences. More precisely, the focus of experience design is on creating new experiences through a deliberate construction of the service circumstances to deepen the engagement and emotional links with users (Shedroff 2001; Pullman and Gross 2004). Metaphorically speaking, designer is a "teller of materially expressed stories" (Usenyuk 2008). Linking this quality with the industry for producing and consuming stories, tales and experiences, gave rise to a new professional field, i.e. "tourism experience design" (a concept formulated by Tussyadiah 2014).

With regard to the research and educational practice of the ADS, the concept of experience design fits perfectly to describe the main approach in tourism-oriented projects. Drawing on key theoretical foundations that frame the practice of experience design in tourism, such as human-centeredness, iterative designing process, and a holistic experience concept (Tussyadiah 2014), the ADS developed its own design approach to the Arctic tourism development based on three key points:

- **Safety/adaptation**: this point is rooted in general human-centeredness of design, with a particular aim to introduce users to a new/extreme environment and new culture gently, as well as to keep them unharmed[2] and healthy during and – as long as possible – after the trip;
- **Interactivity**: this point implies tangible interactions (between tourists and objects that constitute the physical environment of the tourist trip, i.e. special clothing and equipment, transport vehicles, dwelling units), social interactions (between tourists and local communities), and post-trip interactions (between tourists and objects associated with the destinations, i.e. souvenirs);
- **Sustainability**: this point stems from the very mission of ADS, i.e. understanding of and integrating local/indigenous knowledge into the design process, and implies using tourism as a tool "to protect the natural and social capital upon which the industry is built" (McCool and Moisey 2008) and "maintain the Arctic as a desirable place to live" (Miettinen 2012).

Today, across the vast territory of the Russian North, there is a substantial imbalance between what the territory can provide and the tourism industry can utilize. As Maher et al. (2014) observe, local tourism offerings do not vary considerably and are primarily constituted of "hunting and fishing trips, ethnographic tours based on the traditions and culture of the indigenous people of the area, adventure tourism (including snowmobile safaris, white-water rafting, hiking and trekking)".

To rebalance the existing asymmetry between the content of tourism offerings and actual tourism resources of the Russian North, two modes of transforming resources into genuine tourist experiences with the aid of design – static and dynamic – are suggested here (Table 24.1) (terms by Usenyuk 2008).

In the following section, each mode is illustrated by exploring two connected cases: firstly, through a project developing a 'static' museum complex; secondly, through a project creating 'dynamic' lifetime experience in an extreme authentic location. Implications for Arctic tourism principles and practices arising from the presented design approach are discussed after that.

---

[2] In case of the location, physical safety is critical, as Maher et al. (2014) mention the system of emergency rescue as one of the main issues of the tourism industry in the Russian North: it is "yet to be developed and along with the rather specific Russian safety standards (drinking while driving, skepticism towards use of the safety equipment) adds to the uncertainty surrounding tourists' safety in the Arctic."

Table 24.1 Two modes of transforming tourist resources into tourist experiences

	Static	Dynamic
Core idea	Sights are delivered to a tourist	A tourist is delivered to sights
Core experience	To experience the Arctic/North remotely	To experience the Arctic/North proximally, face-to-face
Tourism touch points	Places of deliberate / artificial concentration of tourist value, which are organized in easily accessible areas	Places of natural concentration of tourist value, which are difficult to access
Forms of embodiment	Museums, exhibitions, theme parks	Guided tours, adventurous safaris, long-term excursions, expeditions, etc.
Main task for designers	To create an illusion of immersion by developing an atmosphere provoking tourists into feeling/learning/acquiring new skills, etc.	To create an atmosphere of safety by developing an autonomous system of material objects that support human adaptation to extreme environment in the short term

24.2 Case Studies

24.2.1 Case Study 1: The Mammoth Museum in Lugovskoye Area

The first case was a commissioned design project conducted by the group of design-ers from the ASD together with researchers from the Ethnographic Bureau, Ekaterinburg. The client was the administration of the State Museum of Man and Nature from the city of Khanty-Mansyisk.

The project took place at the Lugovskoye area in Khanty-Mansi Autonomous Okrug, i.e. marshy lowland filled with ancient bones of mammoths.[3] In summer 2007, the Lugovskoye site became a place of creative collaboration between design-ers and historians and archaeologists working on an ambitious project aimed at developing a fundamentally new organizational and visual concept for a "museum in Western Siberia."

A detailed analysis of the situation revealed the distinctive value of both the geo-graphic location and the nature of the resource: at that time (and until now), there were no museums established directly on sites of archeological excavations in Russia. Besides that, Lugovskoye is located next to the city of Khanty-Mansyisk, and a federal automobile road goes in close proximity. Thus, for the planned

---

[3] The timescale of the project: the excavations on site have been conducted over a number of years since 1998. In 2004 the Lugovskoye area was listed as one of the famous regional sights of excep-tional historical and scientific value, and further declared as a natural historical monument on the international scale

The design stage began in 2007. After the successful presentation of the concept, the planning phase was scheduled for the next two years, but lately the whole project was cancelled due to the changed political and economic situation in the Okrug.

museum complex the situation was uniquely arranged to combine natural concentration of tourist value with easy access. However, the value of the archaeological site "as such" was rather unclear for non-professionals and therefore was not enough to attract and satisfy a wide range of tourists. The key issue was the quality of tourist experience that the planned museum complex could provide – whether immersing into Siberian North in a virtual/remote and comfortable way could provide all or the most important of the experiences that visiting the Siberian wilderness in-person could. In this case, the added value was the opportunity to experience not just Siberian North but the North thousands years ago. To make this virtual leap into the distant history, the project team suggested two levels of further design development: organizational/intangible and material/tangible. The organizational level included development of specialized tourist services, ranging from short-term introductory excursions to participatory tours to the archaeological site. To ensure full immersion into the Ice Age, all the tours would include interactive programs to study paleontological findings and ancient technologies. Also, to maintain constant tourist interest to the location, the list of attractions would include: convenient and well-planned infrastructure (e.g. parking lots, public catering, facilities for entertainment and short-term stay, etc.); regularly updated educating and entertaining content of museum programs devoted to the history, biology, geology of the region, as well as to popularization of up-to-date scientific findings; and competitions and other events. Apart from that, the museum complex would include a Mammoth Institute with a permanent paleontological laboratory located on the territory of paleontological excavations, where in-house scientists and visiting researchers would work on collection and further processing of the field data year-round.

To reveal both the context and content of the unique location to users/visitors and to fulfill the above-mentioned organizational needs, the team developed a design concept with emphasis on its outstanding visual appearance, yet aimed to maintain the local nature and culture. While the predominantly flat landscape with the monotonous road leading at the future building did not allow for any dramatic/theatrical scenery, designers decided to put forward the building per se. The exterior of the museum was inspired by a bio-metaphor of a mammoth (Fig. 24.1). Several units divided spatially and functionally constitute an architectural entity by resembling the body of this prehistoric animal lying on its side. From the road, the building appears as a cross-section of permafrost: a massive glass wall with silhouettes of mammoths frozen into it. On the other side facing the direction of the forest, in the buildings resembling tusk fragments designers placed a hotel, observation areas and laboratories.

According to the feedback collected during the project's presentation at the State Museum of Man and Nature of Khanty-Mansyisk, the planned visual appearance of the museum complex, inspired by both the nature and mythology of the location, was equally welcomed by the locals (including indigenous communities) and potential tourists: while the former perceived the building as an organic part of the land but creatively updated, the latter ones noted its exotic but still recognizable look.

Fig. 24.1 The Mammoth Museum: visual presentation. Explanation of the numbers in the panel (*bottom, left*): *1* – exhibition unit; *2* – Mammoth Research Institute; *3* – hosting and catering facilities; *4* – paleontological research lab; *5.1–5.4* – parking lots (Courtesy of D. Kukanov)

This case illuminates the potential of design to augment the naturally given resources and deliver available attractions to tourists in a new – locally rooted, engaging and aesthetically touching – form.

24.2.2   Case Study 2: The Tourist Complex in the Archipelago of Novaya Zemlya

The second case was a master's degree project conducted by a group of seven students in 2012. This project illustrates the arctic application of the 'living lab' concept, described previously.

The archipelago of Novaya Zemlya, Arkhangelsk Oblast, was chosen as a location of the project aimed at developing a "tangible provocation" into a hypothetical model of comfortable human existence in the Far North (Fig. 24.2). The students used a figurative tourist complex as a 'living lab', i.e. a natural-social environment, in which the combination of environmental, social, economic and technological conditions provided for developing and testing objects, technologies and services to facilitate the physical, social and cultural adaptation of different groups of Arctic dwellers and visitors at a community level. To develop their visions of the modeled situation the students worked both in the studio and the field – not on the exact site,

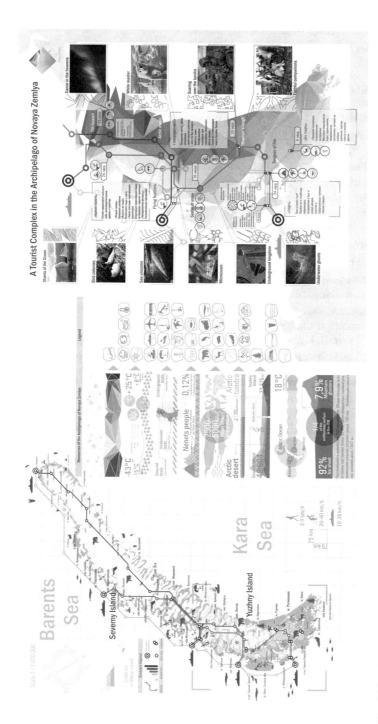

Fig. 24.2 The Archipelago of Novaya Zemlya: the design analysis of tourist resources (Courtesy of: N. Golyzhbina, Yu. Konkova, E. Fadeeva, A. Ufimtseva, I. Putilova, I. Novoselova and E. Shevshenko)

but in alternative locations in Northern and Polar Urals. They were also granted the access to the internal databases and collections of the Institute of History and Archaeology, the Ural branch of Russian Academy of Sciences. As a result, they came up with several design directions to follow:

- *Image-making*: an overall aesthetic appearance, which 'orchestrates' all tangible and intangible components of tourist safety and experience in the extreme environment;
- *Storytelling*: a gradually revealed process of producing and consuming tourist experiences;
- *Material setting*: a system of material objects based on local natural factors, which is used to coordinate a complex interplay between the extreme environment and human beings, i.e. to manage risks and diminish the overall extremity;
- *Mobility*: the combination of physical and mechanized movements, which plays the key role for human adaptation and survival by ensuring the proper functioning of the body. Transportation becomes a "smart carrier," i.e. a partner that not only carries people but also encourages them to switch to other kinds of movement, e.g. walking, running, etc. at various parts of the trip;
- *Relaxation*: a system of physio- and psychological adaptation as "software" that allows creating a unique and secure tourist experience. At the same time, the system of material objects serves as "hardware", i.e. intangible elements of the tourist complex that introduce the visitors to the extreme conditions.

The project yielded the following tangible outcomes (Fig. 24.3): a detailed scenario of events and actions/activities of a tourist during the trip; sets of personal equipment for the identified variety of tourist activities; a set of personal equipment for emergency rescue; a mobile dwelling unit; a chain of personal and communal transport vehicles that facilitates various kinds of tourist mobility; means of physical and psychological relaxation for extreme/adventurous tourist trips; a visual design concept of the tourist complex.

To conclude, in this case study the tourism application area gave rise to a radical revision of existing and development of fundamentally new methods for developing equipment, transport vehicles, housing, etc. The tourism framework revealed the potential of the modeled 'living lab' to co-develop and use a 'playful' system of safety equipment and services as essential elements of tourist experience during adventurous trips. In general, the field of Arctic tourism provided a prospective proving ground for solutions that might have further applications in space and other extreme and isolated environments.

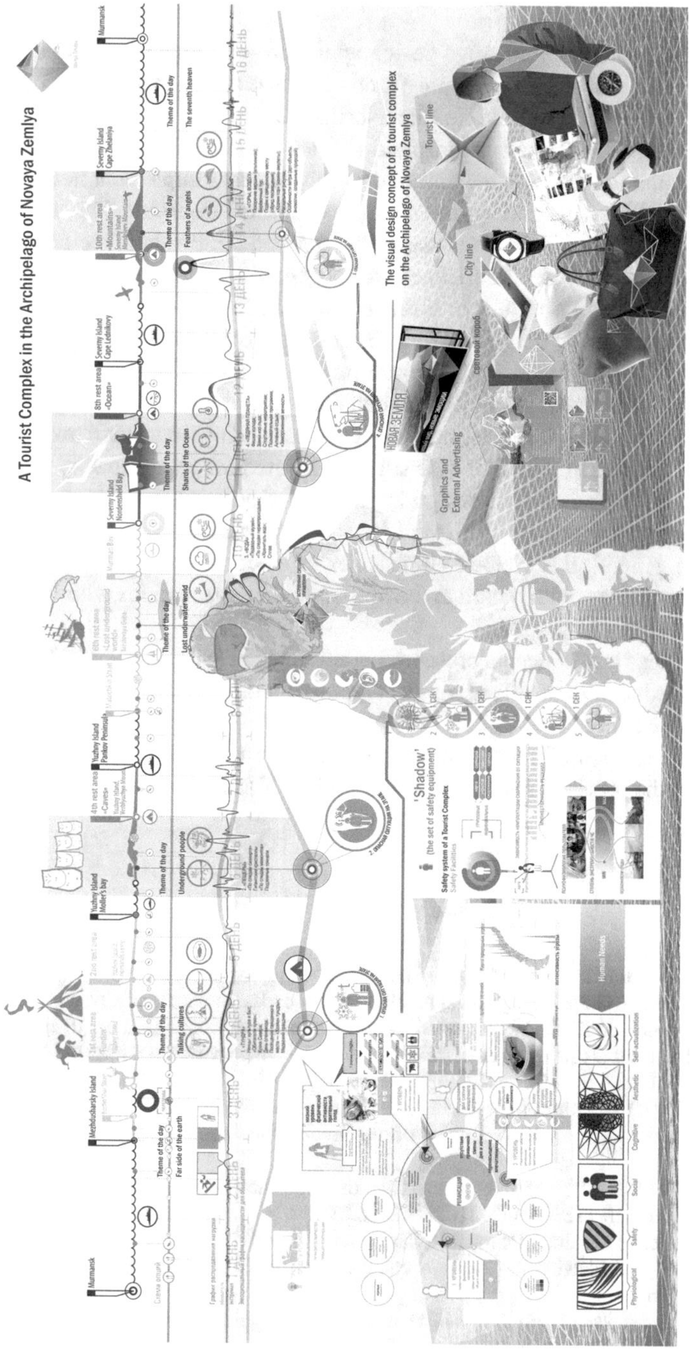

Fig. 24.3 The project outcomes; counterclockwise: a calendar-based scenario linked to the transport chain, the system of relaxation, the set of personal equipment/clothing, the concept of visual design (Courtesy of: N. Golyzhbina, Yu. Konkova, E. Fadeeva, A. Ufimtseva, I. Putilova, I. Novoselova and E. Shevshenko)

24.3 Conclusions

This chapter presents an innovative design approach – originated and tested at the Arctic Design School – for developing arctic tourism at the specific geographical and social-cultural setting of the Russian North.

The Arctic tourism framework inspired the designers of the approach to revisit their professional toolkit with the aim to develop, test and implement new ways of human comfortable living and working in the extreme environment. The 'living lab' approach presented in the second case study can potentially facilitate co-creation between designers and users/tourists as well as enable interaction among relevant Arctic stakeholders: in addition to tourism industry, those from extractive industries, military and a wide range of Arctic-related research and public organizations.

On the other hand, the design approach on the tourism principles and practices existing in the areas under study illuminated the opportunities of shifting the focus from tourism resources and destinations to human beings and their experiences acquired during the trip. The first case study illustrated the joint capacity of artistic imagination and organizational design in creation of genuine experience in a location with no previous tourism. While the focus on experiences is not a new approach in tourism in general, this chapter emphasizes its importance – and importance of human-centeredness – with regard to the tourism linked to locations with vulnerable nature and culture.

Also, the ideas and concepts presented in this chapter will be relevant for design and tourism educators involved in developing methodologies for collaborative research and teaching.

Overall, the chapter offers only a starting point for investigating the fruitful interplay between the tourism and design in the setting of Polar Regions.

References

Arctic Human Development Report (2004) Stefansson Arctic Institute, Akureyri

Bergvall-Kareborn B, Hoist M, Stahlbrost A (2009) Concept design with a living lab approach, in: System Sciences, 2009. HICSS'09. 42nd Hawaii International Conference on. IEEE, pp 1–10

Binkhorst E, Dekker TD (2009) Agenda for co-creation tourism experience research. J Hosp Mark Manag 18:311–327. doi:10.1080/19368620802594193

Brigham LW (2007) Thinking about the Arctic's Future. The Futurist 41(5):27–34

Duhaime G, Caron A (2006) The economy of the circumpolar Arctic. Econ North Oslo No Stat Nor 16–25

Kharlampieva N (2016) Theory and methodology of the Arctic tourism development. Arct North 23:124–129. doi:10.17238/issn2221-2698.2016.23.124

Maher PT, Gelter H, Hillmer-Pegram K, Hovgaard G, Hull J, Jóhannesson GT, Karlsdóttir A, Rantala O, Pashkevich A (2014) Arctic tourism: realities and possibilities. Arct Yearb 2014:290–306

McCool SF, Moisey RN (2008) Tourism, recreation, and sustainability: linking culture and the environment, 2nd edn. CABI, Wallingford/Cambridge, MA

Miettinen S (2012) Service design, radical innovations and arctic wellbeing, in: arctic design – opening the discussion. University of Lapland, pp 28–33

Pullman ME, Gross MA (2004) Ability of Experience Design Elements to Elicit Emotions and Loyalty Behaviors. Decis Sci 35:551–578. doi:10.1111/j.0011-7315.2004.02611.x

Shedroff N (2001) Experience design. New Riders, Indianapolis

Simon HA (1996) The sciences of the artificial, 3rd edn. MIT Press, Cambridge

Slavin SV (1972) Osnovnye voprosy povysheniya effektivnosti, razvitiya i razmescheniya proizvoditel'nykh sil Severa SSSR [Fundamental Issues of Increasing the Efficiency, Developing and Locating the Productive Forces of the USSR's North], in: Northern Issues [Problemy Severa]. Nauka, Moscow

Tussyadiah IP (2014) Toward a theoretical foundation for experience design in tourism. J Travel Res 53:543–564. doi:10.1177/0047287513513172

Usenyuk S (2008) The Russian North: Northern tourism through design professionals. Tour Hosp Plann Dev 5:131–147. doi:10.1080/14790530802252792

Part V
Arctic Safety

Chapter 25
Maritime Operations and Emergency Preparedness in the Arctic–Competence Standards for Search and Rescue Operations Contingencies in Polar Waters

Johannes Schmied, Odd Jarl Borch, Ensieh Kheiri Pileh Roud,
Tor Einar Berg, Kay Fjørtoft, Ørjan Selvik, and James R. Parsons

Abstract Emergencies on large passenger ships in the remote High North may lead to a mass rescue operation with a heavy strain on the emergency preparedness systems of the Arctic countries. This study focuses on the need for competencies related to large-scale Search and Rescue operations (SAR operations) amongst the shipping companies, vessels and governments involved. A SAR operation is the activity related to finding and rescuing people in distress. Several international standards, in particular the conventions by the International Maritime Organization (IMO), provide direction for education and training of seafarers and rescue staff. This study elaborates on the operational competence requirements for key personnel involved in large scale SAR operations. Findings from real SAR incidents and exercises provide in-depth understanding on the operational challenges. The chapter gives directions for competence programs, beyond obligatory international standards, and recommendations for further research.

J. Schmied (✉) • O.J. Borch • E.K.P. Roud
Nord University, Bodø, Norway
e-mail: johannes.schmied@nord.no; Odd.j.borch@nord.no; Ensieh.k.roud@nord.no

T.E. Berg • K. Fjørtoft • Ø. Selvik
SINTEF Ocean, Trondheim, Norway
e-mail: Tor.berg@sintef.no; kay.fjortoft@sintef.no; Orjan.selvik@sintef.no

J.R. Parsons
Memorial University of Newfoundland, St. John's, Canada
e-mail: jim.parsons@mi.mun.ca

© The Author(s) 2017
K. Latola, H. Savela (eds.), *The Interconnected Arctic — UArctic Congress 2016*,
Springer Polar Sciences, DOI 10.1007/978-3-319-57532-2_25

25.1 Introduction

The Arctic maritime regions are characterized by unique and challenging conditions including harsh weather, cold climate, remoteness from harbors and other infrastructure, and a vulnerable environment (Borch and Batalden 2015; Gudmestad et al. 1999; Løset et al. 1999). Incidents involving vessels with many people on board are among the most challenging emergencies to prepare for and to prevent (Ho 2010; Marchenko et al. 2015; Lasserre and Pelletier 2011).

Regional authorities in several Arctic areas including the Northern Sea Route, the North West Passage, and the Spitsbergen region may experience severe capacity and competence challenges in case of an accident involving passenger ships, due to the larger passenger numbers on the vessels and the limited resources available (Arctic Council 2009; Johnston et al. 2012; Roud et al. 2016). An emergency on a cruise ship may have severe consequences (Lois et al. 2004; Vanem and Skjong 2004) as highlighted in the Costa Concordia accident. Several related key challenges may emerge including SAR, fire-fighting, oil recovery operations and vessel salvage.

Regulations such as the International Code for Ships Operating in Polar Waters (Polar Code) may reduce the probability of large scale accidents and increase preparedness on board the vessels (Jensen 2016; Bai 2015). Governments may demand special competence and pilots onboard, and also restrict access to challenging sea regions.

However, there is a risk for severe accidents (Marchenko et al. 2015). Regional studies have shown limitations in SAR capacities, including the SAR technology adapted to the region. This includes information and communication tools for remote areas such as Single Window (Fjortoft et al. 2011), innovation in Arctic vessel and rescue equipment (Berg et al. 2013; Gudmestad and Karunakaran 2012; Torheim and Gudmestad 2011) and development of Arctic infrastructure (Dodds 2013). There is a need for operational knowledge, especially on cooperation among the broad range of actors involved in major accidents, including the crew of the vessel in distress, joint rescue coordination centers (JRCC), coast guard, private preparedness organizations, Samaritan vessels, police, the ship owners and their stakeholders (Borch and Andreassen 2015).

Differences in organization and management principles of stakeholders within emergency networks may hamper partnership and cooperation across institutions and borders. Recently, the Incident Command System (ICS) has gained importance in the response management structure of Arctic states such as the USA, Canada and Norway (Bigley and Roberts 2001). However, other actors such as the NATO military organizations apply other systems with a different management structure. For example, in Norway the police and the military apply their management structure, while the fire and rescue brigades and oil recovery authorities have introduced ICS into their operations (The Roksund Government Committee 2016).

In this respect, there is a need to look closer into the actual roles and positions of operators along the SAR value chain. As an example, the supportive organization and the competence of the on-scene commanders have not been thoroughly studied in the context of Arctic remoteness and mass rescue scenarios (Borch and Andreassen 2015; Małyszko and Wielgosz 2016).

This chapter discusses how the competence of key Arctic maritime SAR personnel needs to be developed in order to manage incidents involving multiple emergency services with different modes of operation. Competence requirements are elaborated upon, both at the operational coordination and command level as well as at the local on-scene tactical level.

Only a few studies have focused on the role of maritime SAR-leaders. Crichton et al. (2005) reflect on incident command skills on oil platforms with respect to the five categories of situational awareness, decision-making, teamwork, leadership and communication. Borch and Andreassen (2015) emphasize the roles of emergency preparedness managers.

Creating common situational awareness is the critical first part of any SAR operation. Knowing what is going on is one of the key prerequisites in crisis response (Oomes 2004; Endsley and Garland 2000). However, areas like the Arctic are prone to limited information and situational awareness due to limited communication facilities (Behlke 2013), lack of vessels and equipment, as well as personnel (Berg et al. 2013; Borch and Andreassen 2015; Rottem 2014). The capability to make decisions based on limited knowledge may improve through education. Klein (1993), Orasanu and Connolly (1993), Cosgrave (1996) and Dreyfus and Dreyfus (1986) claim that the ability to assess situations can be improved through practiced understanding – in other words exercises and training.

Decision making in a SAR value chain is a difficult command skill and characterized by uncertainty about the cause-consequence links, due to limited information and knowledge as to both context and SAR-tools available (Liu et al. 2014). One aspect, which is particularly highlighted in recognition-primed decision model and naturalistic decision-making (Liu et al. 2014), is the importance of previous events or training and exercises, which enforces the importance of having experienced and suitably trained decision makers. According to both the incident command approach by Crichton et al. (2005) and Managerial Roles by Mintzberg (2009), a broad range of leadership tasks plays a central role in managing demanding situations.

Due to the scale of a mass rescue operation compared with the lack of immediate resources available, close cooperation and teamwork amongst several organizations is needed. Trust between the organizations is important. Sako (1998) suggests that trust can be based on the trustee establishing competence, goodwill or contractual promise keeping. The trustor on the other hand becomes vulnerable to the trustee's actions (see Wilson et al. 2007).

Specifically, contractual trust may include standard operation procedures (SOP) which frame teamwork (Ivanova 2011; Ivanova and Sydnes 2010; Sydnes and Sydnes 2013). A continuous information flow via SOPs may enable more efficient use of resources, coordination and lower risk of operations, possibly at all levels from the operational scene up to the national government level. Interpersonal communication hereby plays a major role (Mintzberg 2009).

Finally, competence based on communication and communication systems is highly important and needs to be established prior to potential incidents (Crichton et al. 2005; Kapucu et al. 2010). Communication may already be compulsory due to pre-existing control mechanisms (Mintzberg 2009). Yet, aspects such as lack of trust due to cultural or political reasons or different language, be it due to different standards in different institutions or different nationalities, may be critical aspects of communication (Comfort and Kapucu 2006; Kapucu 2005; Robinson et al. 2015). In a case study of a SAR-exercise focusing on shared situational awareness and communication, Seppänen et al. (2013) discovered that *"information gaps, the lack of fluent communication, and the fact that there is no common operational picture"* hamper emergency management.

SAR operations at sea are regulated by, among others, the International Convention on Maritime Search and Rescue (SAR), International Aeronautical and Maritime Search and Rescue (IAMSAR)-Manuals 1–3 and the International Convention for the Safety of Life at Sea (SOLAS). SOLAS includes the relevant International Ship and Port Facility Security Code (ISPS-Code). Additional regulatory framework includes the International Convention on Standards of Training, Certification and Watchkeeping for Seafarers (STCW) with the Manila amendments of 2010, the International Convention on Standards of Training, Certification and Watchkeeping for Fishing Vessel Personnel (STCW-F), as well as other IMO conventions with indirect relevance to SAR and standards by standardization societies. With respect to cross-boundary coordination and host nation support (HNS) the United Nations Law of the Sea (UNCLOS) and regional and bilateral agreements on responsibility, rights and exercises are important.

A study by Ghosh and Rubly (2015) supports the importance of the Polar Code, pointing out that emergency management personnel including the on-scene staff such as captains, officers and crew, need to understand and be able to handle potential hazards and risks. Sander et al. (2015) point out the need for further regulations. Even though the Polar Code provides regulations to a certain extent, Jensen (2016) questions the potential power of the Polar Code due to its openness to national interpretation with regard to implementation standards.

The IAMSAR Manuals define the on-scene coordinator (OSC) as the most capable and trained leading person on site (IMO 2016). According to Rake and Njå (2009), the OSC needs to be well-rendered, aware of the situation, a good communicator, quick at information processing, a quick decision maker and swift at improvising. The OSC has to be clear, available, cooperative and lead with authority. Also the OSC has to have best suitable equipment and support from the JRCC (see Małyszko and Wielgosz 2016; Ansell et al. 2010).

The IAMSAR manuals of 2016 for the first time also establish the Aircraft Coordinator (ACO) in an equally important manner as the OSC (IMO 2016). The ACO should maintain high flight safety, advise and support as well as increase effectiveness of the operation (Ibid; USCG 2010). Few systematic studies related to marine ACO competence have been found.

A facilitating Command System is also essential for successful emergency response. It is traditionally structured hierarchically and has clear command and control agreements (Owen et al. 2015). Particularly ICS are designed for large-scale operations and support coordination, flow of information, best possible use for on-scene personnel and timely decision making (Rimstad et al. 2014; Boersma et al. 2014).

Rimstad et al. (2014) claim that crises should be managed at the lowest possible level, procedures should be every day procedures, and both public and private actors should jointly provide all resources available. Yet, there is a need for mission commanders to coordinate and facilitate cooperation beyond organizations and borders (Politidirektoratet 2011). Christensen et al. (2012) show that information flow and mobilization should run as automatically and efficiently as possible and observe increased vertical information flows within an organization, before information is shared horizontally.

By keeping the complexity and uncertainty of Arctic maritime emergency operations in mind, there is a special challenge related to the role of the emergency system managers and their command structure. There is a need to develop a look into criteria for relevant command skills and competences for on-scene personnel and the command system, and to find the tools needed for education and training of key personnel along the whole SAR-value chain.

25.2   Methods

This study followed an in-depth, qualitative research strategy focusing on command structures, managerial roles and the emergency management competence needs. Illustrative cases were chosen to highlight the complexity and challenges of SAR operations in Arctic waters.

Data were collected from observations both on full scale and table top exercises (TTX). The observational studies were followed up by unstructured interviews with key personnel after the SAR exercises. In addition, a combination of interviews, studies of logs and investigation reports from major accidents were included.

The study follows up on eleven cases (Table 25.1) taken from a larger research initiative with 27 cases. Both domestic and international exercises were included in the study to emphasize the challenges in organizational- and cross-border cooperation. Related standards and regulations in the Arctic were reviewed and reflected upon according to data from exercise observations. In the analysis, the data from exercises and incidents were screened on competence gaps and challenges in order to determine further necessary activity to develop relevant command skills and competences for on-scene personnel and the command system.

25.3   Results and Discussion

In a mass rescue operation the number of persons in the water, in rafts and in life boats may be overwhelming. There may also still be people on board the vessel in distress. Real incidents like the Maxim Gorkiy collision with ice in the Spitsbergen region and the Costa Concordia grounding in Italy showed that a major challenge in mass rescue incidents is to perform "vessel or rescue unit triage" i.e. prioritize who

Table 25.1 List of exercises and incidents presented in this study

Name	Exercise/incident/other	Year	Month	Place	Documents
Maxim Gorkiy	Incident	1989	June	Svalbard	Logs, reports, articles, case study, interview
Costa Concordia	Incident	2012	January	Italy	Report
KV SAR 2016	Newsletter on exercises and incidents	2015/16	Several	Norway	Summary
Barents Rescue exercise	Full scale exercise	2015	September	Finland	Observation
Exercise Barents	TTX/Full scale exercise	2015/16	Several	Barents	Observation, protocol, brief
Exercise Helgeland	Full scale exercise	2015	May	Norway	Log book, observation, protocol
Exercise Nord	Full scale exercise	2015/16	April	Norway	Log book
TTX KV Sortland	Table top exercise	2016	April	Norway	Brief, log-book
SARex	Full scale exercise about Polar Code	2016	April	Svalbard	Observation, log books, reports
Arctic SAR	Meta table top exercise	2016	April	Video Conference	Observation, protocol, summary
AECO SAR workshop	Table top exercise	2016	April	Iceland	Report, observation

to rescue based on limited information. This is especially a challenge for on-scene personnel. It also lays a heavy burden on the mission coordinator at the JRCC, who may experience challenges to allocate adequate resources fast enough.

In the Arctic, helicopter resources may play a vital role especially in an early phase. The Air Coordinator role is also reported to become more demanding in mass rescue incidents and needs further exercise and training with respect to the IAMSAR manuals (see IMO 2016). Exercise Nord and Arctic SAR indicated that airborne professionals and SAR non-professional captains, officers and masters are often first on-site and therefore need to be well-educated for the OSC-role (see Klein 1993; Orasanu and Connolly 1993; Cosgrave 1996; Dreyfus and Dreyfus 1986).

Therefore, more specific training and development for potential ACOs and OSCs is in demand. In the TTX KV Sortland it was discussed to assign this task to a person that could also fill the figurehead-role, as described by Mintzberg (2009).

The importance of crew situational awareness, spontaneous adaptation to changes and knowledge on procedures, standards and automated processes were highlighted by the reports of the Costa Concordia incident. In this case, crew training on passenger management would have been particularly beneficial to prevent casualties. The Polar Code's chapter "Manning and Training" discusses these tasks, albeit on a rather general level. As a consequence, the IMO is working on stricter procedures in this respect.

For the master of the distress vessel and the on-scene coordinator, knowledge gathering and operational planning for the steps ahead including improvisation was important, but seldom trained at a realistic scale (source: Exercise SARex and Exercise Nord). Periodic dialogue and data sharing as well as training with the SAR operators and the JRCC on resource allocation and priorities are critical in this respect.

Experience from Exercise Barents 2015 showed that in some cases, difficult political situations or heavily bureaucratic bilateral cooperation is reducing the effects of the exercise. Knowledge and testing of the underlying Arctic SAR control mechanisms are recommended for improvement. Further efforts to increase awareness and competence in the command system in terms of cross border cooperation and HNS may include a *"written brief on the structure of the host nation's command system"*, as recommended from the Barents Rescue exercise 2015.

Preparation of ships before they go to the Arctic may help captains in case of emergencies. Not the least, the importance of fast alarm is central. The Costa Concordia and Maxim Gorkiy incidents are two examples where alarms came more than an hour after the accident. Furthermore, harbors and their crew need to adapt to increased ship sizes and activities and be prepared for taking care of a large number of rescued persons (Haugen and Fjortoft 2011). Decision support systems, as suggested by Małyszko and Wielgosz (2016), may help to determine methods for managing all these data and make decisions under uncertainty. Also, the establishment and preparation of action plans helped both the OSC as well as the whole command system to perform efficiently (Report from Exercise Helgeland on oil spill recovery).

In this respect, it is vital to know and efficiently communicate where the responsibilities of each person (OSC, ACO, Captain, Samaritan Vessels, etc.) start and end. Taking care of OSC responsibilities may demand several persons, and how to choose them is not elaborated upon in the IAMSAR manuals (IAMSAR Volume II). Actors should be aware of differences and regulations of the control mechanisms and, in turn, commit to trust, as outlined by Sako (1998).

Regardless of the field of research, researchers agree on the importance of trust in social interaction (Rousseau et al. 1998). Data from domestic exercises and the AECO SAR workshop showed that those who had developed relations prior to the actual emergency response through daily collaboration had built trust, and thus experienced fewer challenges in organizational collaboration on complex tasks. This includes willingness to collaborate, information sharing and shared values or standards (see Kapucu 2006).

Preparedness of the vessel's communication lines and competence in use is crucial. Connection capacities may be limited in the Arctic, as discussed previously, causing decision-making under ambiguity (Behlke 2013). Preparations and installations before a cruise journey (e.g., via Polar Water Operational Manual), including actual learning from the experiences of previous cruises is therefore important.

In order to support communication, chat systems and shared written communication logs have proven to be an efficient tool for command systems to overcome misunderstandings resulting from oral communication. However, as was indicated in the Norwegian Coast Guard SAR info KV 2016, it is at times still problematic with different chat-programs, different cultures and un-coordinated systems, especially in terms of larger incidents including several command structures or cross-border response. More communication capacity has therefore been requested during exercises between the OSC and HNS resources.

25.4   Conclusions

The objective of this study was to provide insights into the roles of the emergency command system. Arctic SAR operations and especially mass rescue operations are very complex and require enhanced experience from the key personnel. The Polar Code will contribute to increased crew and operator knowledge on the unique conditions of the Polar regions, including what is needed as to safety training and manning for emergency situations. Efficient communication and close interaction between the vessel in distress, the OSCs and the SAR mission coordinators are vital, and should be emphasized in all procedures.

This study has shown that exercises are often too limited in scale and scope and do not illuminate the full range of challenges related to mass rescue operations in the Arctic. Advanced OSC courses should be available for a broader spectrum of personnel including the masters and officers of all vessels in Polar regions who are often the first on site. For vessel captains, additional course modules in the Global Maritime Distress and Safety System course on OSC/ACO-air coordinator roles in the Arctic need to be developed as well as integrated in the Polar water operation manuals and contingency plans. This is particularly important for the crew of large passenger vessels operating in remote areas.

The results of this study indicate that immediate alarm notification and creation of a common situational awareness are key issues induced by the operational knowledge. In mass rescue incidents the leading personnel both on board and on shore, including the OSC, ACO, SAR mission coordinator and staff need to be aware of the capabilities and capacities that may be mobilized in different phases of the emergency. More efforts towards competence sharing are recommended in all stages of the SAR value chain. The operators on different levels need in-depth education on decision making under uncertainty, teamwork and leadership. Cross-institutional cooperation in education may help to pinpoint the bottlenecks within the command systems and reveal opportunities on how to integrate different systems.

Further studies on the coordination of cross-institutional cooperation, managerial roles, competence sharing and the trust aspect may contribute to increased efficiency in joint actions within the Arctic SAR value chain.

References

Ansell C, Boin A, Keller A (2010) Managing transboundary crises: Identifying the building blocks of an effective response system. J Conting Crisis Manag 18(4):195–207

Arctic Council (2009) Arctic marine shipping assessment (AMSA). PAME and Arctic Council, Oslo

Bai J (2015) The IMO Polar Code: The emerging rules of Arctic Shipping Governance. Int J Mar Coast Law 30(4):674–699

Behlke R 2013 Maritime user requirements on navigation and communication solutions at high latitudes (Svalbard)-with an outlook on the MARENOR project. In: AGU Fall Meeting Abstracts. p 2242

Berg TE, Holte EA, Ose GO, Færevik H (2013) Safety at Sea: Improving Search and Rescue (SAR) Operations in the Barents Sea. In: ASME 2013 32nd International Conference on Ocean, Offshore and Arctic Engineering. American Society of Mechanical Engineers, pp V006T007A007-V006T007A007

Bigley GA, Roberts KH (2001) The incident command system: High-reliability organizing for complex and volatile task environments. Acad Manag J 44(6):1281–1299

Boersma K, Comfort L, Groenendaal J, Wolbers J (2014) Editorial: Incident command systems: A dynamic tension among goals, rules and practice. J Conting Crisis Manag 22(1):1–4

Borch OJ, Andreassen N (2015) Joint-task force management in cross-border emergency response. Managerial roles and structuring mechanisms in high complexity-high volatility environments. information, communication and environment: marine navigation and safety of sea transportation:217

Borch OJ, Batalden B-M (2015) Business-process management in high-turbulence environments: the case of the offshore service vessel industry. Marit Policy Manag 42(5):481–498

Christensen T, Lægreid P, Rykkja LH (2012) How to cope with a terrorist attack?–a challenge for the political and administrative leadership. COCOPS working paper 6

Comfort LK, Kapucu N (2006) Inter-organizational coordination in extreme events: The World Trade Center attacks, September 11, 2001. Nat Hazards 39(2):309–327

Cosgrave J (1996) Decision making in emergencies. Disaster Prev Manag 5(4):28–35

Crichton MT, Lauche K, Flin R (2005) Incident command skills in the management of an oil industry drilling incident: A case study. J Conting Crisis Manag 13(3):116–128

Dodds KJ (2013) Anticipating the Arctic and the Arctic Council: pre-emption, precaution and preparedness. Polar Record 49(02):193–203

Dreyfus H, Dreyfus S (1986) Mind over machines. Mind over machines

Endsley MR, Garland D (2000) Theoretical underpinnings of situation awareness: A critical review. Situation awareness analysis and measurement:3–32

Fjortoft K, Hagaseth M, Lambrou M, Baltzersen P, Papachristos D, Nikitakos N (2011) Maritime transport single windows: issues and prospects. Transport systems and processes: Marine navigation and safety of sea transportation:19

Ghosh S, Rubly C (2015) The emergence of Arctic shipping: issues, threats, costs, and risk-mitigating strategies of the Polar Code. Aust J Maritime and Ocean Aff 7(3):171–182

Gudmestad O, Zolotukhin A, Ermakov A, Jakobsen R, Michtchenko I, Vovk V, Løset S, Shkhinek K (1999) Basics of offshore petroleum engineering and development of marine facilities. Oil and Gas Printing House, Moscow, pp 176–196

Gudmestad OT, Karunakaran D (2012) Challenges faced by the marine contractors working in western and southern Barents Sea. In: OTC Arctic technology conference. Offshore Technology Conference

Haugen T, Fjortoft K (2011) Visions and future needs in Arctic Maritime Operations

Ho J (2010) The implications of Arctic sea ice decline on shipping. Mar Policy 34(3):713–715

IMO (2016) IAMSAR Manual II – Mission Co-ordination. ISBN: 978-92-801-1640-3

Ivanova M (2011) Oil spill emergency preparedness in the Russian Arctic: a study of the Murmansk region. Polar Res 30

Ivanova M, Sydnes AK (2010) Interorganizational coordination in oil spill emergency response: a case study of the Murmansk region of northwest Russia. Polar Geogr 33(3–4):139–164

Jensen Ø (2016) The international code for ships operating in polar waters: finalization, adoption and law of the sea implications. Arctic Rev on Law and Politics 7(1):60–82

Johnston A, Johnston M, Stewart E, Dawson J, Lemelin H (2012) Perspectives of decision makers and regulators on climate change and adaptation in expedition cruise ship tourism in Nunavut. Northern Review (35)

Kapucu N (2005) Interorganizational coordination in dynamic context: Networks in emergency response management. Connections 26(2):33–48

Kapucu N (2006) Interagency communication networks during emergencies boundary spanners in multiagency coordination. Am Rev Public Adm 36(2):207–225

Kapucu N, Arslan T, Demiroz F (2010) Collaborative emergency management and national emergency management network. Disaster Prevention and Management: An International Journal 19(4):452–468

Klein GA (1993) A recognition-primed decision (RPD) model of rapid decision making. Ablex Publishing Corporation, New York

Lasserre F, Pelletier S (2011) Polar super seaways? Maritime transport in the Arctic: an analysis of shipowners' intentions. J Transp Geogr 19(6):1465–1473

Liu Y, Fan Z-P, Yuan Y, Li H (2014) A FTA-based method for risk decision-making in emergency response. Comput Oper Res 42:49–57

Lois P, Wang J, Wall A, Ruxton T (2004) Formal safety assessment of cruise ships. Tour Manag 25(2009):93–109

Løset S, Shkhinek K, Gudmestad O, Strass P, Michalenko E, Frederking R, Kärnä T (1999) Comparison of the physical environment of some Arctic seas. Cold Reg Sci Technol 29(3):201–214

Małyszko M, Wielgosz M (2016) Decision support systems in search, rescue and salvage operations at sea. Sci J Maritime Univ Szczecin 45(45):191–195

Marchenko NA, Borch OJ, Markov SV, Andreassen N (2015) Maritime activity in the high North – the range of unwanted incidents and risk patterns. Proceedings of the 23rd international conference on port and ocean engineering under arctic conditions (June 14–18, 2016)

Mintzberg H (2009) Managing. Berrett-Koehler Publishers, San Francisco

Oomes A (2004) Organization awareness in crisis management. In: Proceedings of the international workshop on information systems on crisis response and management (ISCRAM)

Orasanu J, Connolly T (1993) The reinvention of decision making

Owen C, Scott C, Adams R, Parsons D (2015) Leadership in crisis: developing beyond command and control. Aust J Emerg Manag 30(3):15

Politidirektoratet (2011) PBS I – Politiets BEREDSKAPSSYSTEM del I – Retningslinjer for politiets beredskap. POD-publikasjon nr. 2011/04

Rake EL, Njå O (2009) Perceptions and performances of experienced incident commanders. J Risk Res 12(5):665–685

Rimstad R, Njå O, Rake EL, Braut GS (2014) Incident Command and Information Flows in a Large-Scale Emergency Operation. J Conting Crisis Manag 22(1):29–38

Robinson JJ, Maddock J, Starbird K (2015) Examining the role of human and technical infrastructure during emergency response. In: Information systems for crisis response and management, ISCRAM

Rottem SV (2014) The Arctic Council and the search and rescue agreement: the case of Norway. Polar Record 50(03):284–292

Roud EKP, Borch OJ, Jakobsen U, Marchenko N (2016) Maritime emergency management capabilities in the Arctic. In: Proceedings of the twenty-sixth (2016) international ocean and polar engineering conference. International Society of Offshore & Polar Engineers

Rousseau DM, Sitkin SB, Burt RS, Camerer C (1998) Not so different after all: A cross-discipline view of trust. Acad Manag Rev 23(3):393–404

Sako M (1998) Does trust improve business performance. Organizational trust: A Reader:88–117

Sander G, Gille J, Stępień A, Koivurova T, Thomas J, Gascard J-C, Justus D (2015) Changes in Arctic maritime transport. In: The changing arctic and the European union. Brill, p 81–114

Seppänen H, Mäkelä J, Luokkala P, Virrantaus K (2013) Developing shared situational awareness for emergency management. Saf Sci 55:1–9

Sydnes AK, Sydnes M (2013) Norwegian–Russian cooperation on oil-spill response in the Barents Sea. Mar Policy 39:257–264

The Roksund Government Committee (2016) The Roksund working group on armed forces support to the police (in Norwegian: Forsvarets Bistand til Politiet – Rapport fra arbeidsgruppen for utarbeiding av forslag til ny bistandsinstruks). ISBN: 978-82-7924-088-4, vol 09/2016 – opplag 100. Norwegian Ministry of Justice and Public Security, Available at: https://www.regjeringen.no/globalassets/departementene/fd/dokumenter/rapporter-og-regelverk/20160930-rapport-fra-arbeidsgruppen-for-utarbeiding-av-forslag.pdf

Torheim S, Gudmestad OT (2011) Secure launch of lifeboats in cold climate: looking into requirements for winterization. In: ASME 2011 30th international conference on ocean, offshore and arctic engineering. American Society of Mechanical Engineers, pp 931–939

USCG (2010) International manual for aircraft coordinator. Available at: https://www.uscg.mil/hq/cg5/cg534/nsarc/BalticAcoManual.pdf

Vanem E, Skjong R (2004) Collision and grounding of passenger ships–risk assessment and emergency evacuations. In: Third international conference on collision and grounding of ships (ICCGS). p 202

Wilson KA, Salas E, Priest HA, Andrews D (2007) Errors in the heat of battle: Taking a closer look at shared cognition breakdowns through teamwork. J Human Factors Ergon Soc 49(2):243–256

Chapter 26
Risk Reduction as a Result of Implementation of the Functional Based IMO Polar Code in the Arctic Cruise Industry

Knut Espen Solberg, Robert Brown, Eirik Skogvoll, and Ove Tobias Gudmestad

Abstract The IMO Polar Code states that equipment and systems providing survival support for passengers/crew should have adequate thermal protection for a minimum of 5 days. Based on participant workshops where suppliers, regulators, users and academia were present, the following three functionality requirements were identified as essential for survival: Maintaining cognitive abilities; No uncontrollable body shivering and Functionality of extremities.

Following the participant workshops, a field trial was conducted in Wood Fjord, Northern Svalbard, during the last week of April 2016. The goal of the trial was to identify the gaps in functionality provided by life-saving equipment currently approved by SOLAS and the functionality required to comply with the minimum requirement of 5 days survival, according to the IMO Polar Code.

The trial demonstrated that when utilizing standard SOLAS approved equipment, compliance with the functional Polar Code requirement of protection from hypothermia cannot be expected beyond 24 h of exposure.

K.E. Solberg (✉) • O.T. Gudmestad
Mechanical and Structural Engineering and Materials Science, Faculty of Science and Technology, University of Stavanger, Stavanger, Norway
e-mail: Knut.Espen.Solberg@gmc.no

R. Brown
Ocean Safety Research, Marine Institute, Memorial University, St. John's, Canada

E. Skogvoll
Faculty of Medicine, Norwegian University of Science and Technology (NTNU), Trondheim, Norway

© The Author(s) 2017
K. Latola, H. Savela (eds.), *The Interconnected Arctic — UArctic Congress 2016*,
Springer Polar Sciences, DOI 10.1007/978-3-319-57532-2_26

26.1 Introduction

Cruise ship activity in polar regions has increased in recent years and the trend is expected to continue. With the successful transit of the Crystal Serenity through the Northwest Passage in 2016, there are currently several expedition cruise vessels being commissioned. The increase in the cruise ship industry is also expected to take place around the Svalbard island (Brunvoll 2015).

The International Code for Ships Operating in Polar Waters (The IMO; International Maritime Organization Polar Code) is a supplement to existing IMO instruments, and the intention is to mitigate the additional risks present for people and environment when operating vessels in polar waters (International Maritime Oranization 2016). The code enters into force on 01.01.2017 for newbuilds, and on 01.01.2018 for existing vessels.

Contrary to most of the existing IMO instruments, the International Code for Ships Operating in Polar Waters provides a risk-based approach (ABS 2016) to regulating activity in this area. This means that marine operators are to identify risks and mitigate them through a holistic approach.

According to *IMO Polar Code, Chapter 8 – Life-saving appliances and arrangements*, the life-saving equipment is to provide adequate functionality to ensure human survival for a minimum of 5 days for the anticipated weather conditions (cold and wind) and potential for immersion in polar water.

In an effort to better understand the performance requirements for polar survival equipment, a set of field trials was undertaken with human participants in Wood Fjord, Northern Svalbard in the last week of April 2016.

The goal of the field trials was to identify the gaps in functionality provided by regular SOLAS approved life-saving equipment and the functionality required to comply with the minimum requirement of 5 days survival, according to the IMO Polar Code (Solberg et al. 2016).

26.2 Methods

Two life saving appliances (LSAs) were deployed to the water surface – a 25 person life raft and 50 person lifeboat with 19 and 18 participants, respectively. The participants were mainly personnel from the Coast Guard. The majority of the participants were young men in their early 20s. Due to their training from the Coast Guard, they were accustomed to cold climate conditions and were in general physically fit (completed a 3000 m run in less than 15 min).

All participants wore long woolen underwear under regular shirts and pants. The participants were equipped with different types of SOLAS approved personal protective equipment (PPE). The following gear was utilized:

Neoprene survival suit – Neoprene survival suit with integrated soles, 4 pieces.
Insulated survival suit – Insulated survival suit with integrated soles, 6 pieces.

Non-insulated survival suit – Non-insulated survival suit with integrated soles, 5 pieces.

Thermal protection vest – Standard SOLAS approved thermal protection vest/aid, 6 pieces.

Kampvest with bag – The standard life jacket utilized by the Norwegian Coast Guard. The participants stayed inside a plastic bag (TPA-Thermal Protection Aid), 6 pieces.

Kampvest without bag – The standard life jacket utilized by the Norwegian Coast Guard, 4 pieces.

Nordkapp drakt – The offshore working suit utilized by the Norwegian Coast Guard. The suit with integrated steel toe boots, and loose neoprene gloves, 2 pieces.

Survival suit 307 – The standard survival suit utilized by the Norwegian Coast Guard with integrated soles, 2 pieces.

The participants were constantly monitored by medical personnel and were omitted from the trial when any of the following predefined criteria were met:

- Loss of cognitive abilities
- Loss of body control (uncontrollable shivering)
- Loss of functionality of body extremities

When the participants commenced the exercise, they were warm and dry. There was no water present in the rescue crafts on commencement of the exercise. Introducing water inside the rescue craft would significantly have reduced the participants' survival time (DuCharme 2007).

During the exercise, body core temperatures were monitored and recorded for selected participants. All participants went through a medical examination immediately after aborting the exercise, where cognitive abilities, functionality and body temperature were assessed and documented.

26.2.1 Exercise Validity

The intent of the exercise was to simulate a cruise ship incident during the cruising season in Svalbard. The following boundary conditions were observed:

- Average ambient air temp = −9 °C
- Average wind speed = 2 m/s
- Water temperature = −1.2 °C
- Participant health = above average
- Participant insulation layer = average
- Additional stress factors = marginal

A higher wind speed would be expected to reduce the survival times considerably and the weather conditions observed should be regarded as a "best case".

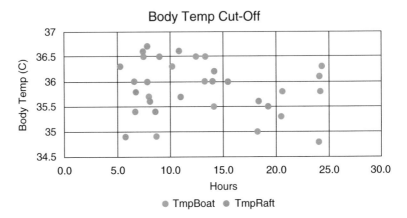

Fig. 26.1 Body Temp Cut-Off – the individual body temperatures at time of abandoning the exercise

As the participants were, on average, not only younger but also fitter than the average cruise ship passenger, the participants' physical condition gave them a higher probability for survival.

The abortion criteria gave a consistent cut-off point for participants, with all aborting the exercise with a core body temperature between 34.7 °C (mild hypothermia) and 36.7 °C (normal) (Fig. 26.1).

In a real scenario, most survivors would be very strongly motivated to stay alive and would be expected to survive for an extended period after our abortion criteria were met. It is however unlikely that the majority of the participants would survive for another 4 days, as required by the Polar Code, using equipment currently approved by SOLAS.

26.3   Results and Discussion

Based on the Kaplan-Meier Survival Plot (Fig. 26.2) it was evident that the cooling process started immediately after the exercise commenced. The first participants aborted the exercise from the raft after about 6 h.

Eight hours into the exercise, the engine in the lifeboat was turned off, removing an essential heat source. After this point in time, neither of the LSAs had a heat source, except what was generated by the participants.

In the life raft, the last participants aborted the exercise after 19 h, while several persons remained in the lifeboat after 24 h.

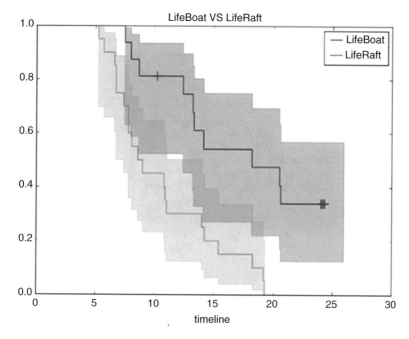

Fig. 26.2 Kaplan-Meier Survival Plot – indicating the fraction of participant survival on the Y-axis and the time spent in the rescue craft in hours on the X-axis (based on abortion criteria)

26.3.1   *Hazard Curve*

The data from the lifeboat plotted as a hazard curve (with confidence interval) shows that the highest hazard was experienced after about 15 h. At around this time, the rate of participants leaving the exercise was at its highest (Fig. 26.3).

The hazard curve for the lifeboat has distinct features: a period of low hazard, a period of increasing hazard and a period of decreasing hazard. For the life raft, the same features could be identified but by the time the life raft reached the survival phase, no participants were left.

The analysis of the hazard curve was broken down into three different phases (Fig. 26.3).

26.3.2   *Stage 1 – Cooling Phase*

During the first 7.5 h, all participants remained in the life-boat. Everyone was well fed, dry and warm prior to entering the rescue craft. In this phase, the participants became accustomed to their situation. During this period, the social structure was established with the lifeboat captain, including a plan on how to distribute resources, e.g. water in addition to distribution of responsibilities, e.g. keeping lookout.

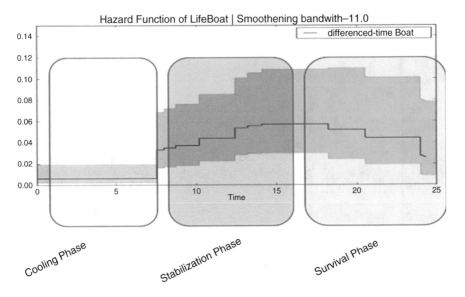

Fig. 26.3 Rescue Craft Phases – an illustration of the different phases, cooling phase (*yellow*), stabilization phase (*red*) and survival phase (*green*), with the timeline (hours) on the X-axis and the Hazard coefficient on the Y-axis for the lifeboat

During this phase, the participants were exposed to the cold natural environment, with an ambient air temperature of about −9 °C and a sea water temperature of −1.2 °C.

26.3.3   Stage 2 – Stabilization Phase

From about 7.5 h into the exercise, participants were starting to abort the exercise. The rate increased steadily until it reached its peak at about 16 h.

Those first to leave were in general participants with only life vests/thermal protective aids. Many of them being wet, typically from condensation inside the rescue craft. The moisture caused an increased heat loss due to evaporative and conductive cooling, which reduces the insulating capabilities of the clothes.

Several also left the exercise early due to significant cooling of their extremities, with the most dominant area of concern being the hands. Cooling of the hands occurred typically because of conducting tasks that required fine motor skills, e.g. opening/closing zippers and opening water bags.

The lifeboat engine was turned off 8 h into the exercise. To increase the internal air temperature, hatches remained closed for the majority of the time. CO_2-level meters showed an alarmingly high CO_2 concentration, and the craft had to be ventilated about every 15 min, depending on the number of participants on board. This process contributed to reducing the interior air temperature. Low O_2-levels also turned out to be a major concern for the participants in the life raft as identified in previous projects (Baker Andrew et al. n.d.).

26.3.4   Stage 3 – Survival Phase

From 16 h onwards, the rate at which participants aborted the exercise slowly decreased until the trial was complete after 24 h. As participants left the rescue craft, space was made available, giving the remaining participants the opportunity to move, generate heat and increase the blood flow to the extremities. The reduced number of persons on board also decreased the need for venting due to increased CO_2 levels.

This far into the exercise the participants were starting to feel fatigue, which resulted in an urge to lie down and rest. Substantial heat loss was experienced from the body parts that were in contact with the cold surfaces inside the rescue crafts. This again resulted in abortion criterion Pt. 2 being met.

By the time the rescue craft had reached Phase 3, the survival phase, the following conditions essential for improving survivability had emerged: Sufficient space to allow movement; Reduced CO_2 levels inside rescue crafts; Established rescue craft routines, giving the participants the ability to predict and remain in "control" of the situation.

26.3.5   Habitable Environment

When a rescue craft is filled with close to 100% of its capacity, the heat generated by the occupants results in a relatively high internal air temperature (Fig. 26.4). From the figure it is also evident that the heat generated by the lifeboat engine adds a significant amount of heat, keeping the internal air temperature stable until it is turned off at about 500 min into the exercise.

The temperature reductions observed at regular intervals for the life raft-curve are a result of the occasions when the participants opened the canopy for venting.

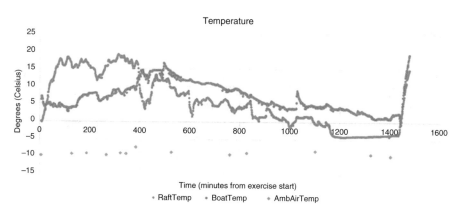

Fig. 26.4 Internal air temperature – the internal air temperature inside the rescue crafts

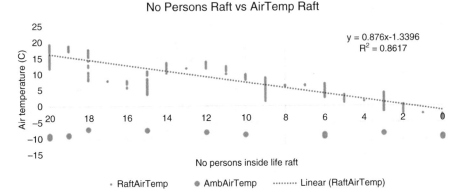

Fig. 26.5 Air Temperature vs. People in the Raft – the internal air temperature is plotted in relation to the number of people inside the life raft

The ambient outside air temperature for the duration of the exercise was relatively steady at between -7 and $-10\ °C$. The decrease in the interior temperature is correlated to the number of persons present inside the rescue craft. This relationship is clearly visible in Fig. 26.5: Air Temperature vs. People in the Raft.

Lack of space resulting in lack of ability to move body limbs was also identified as a major challenge by the participants during the post exercise interviews. The reduced ability to move caused a lack of blood circulation. SOLAS approved LSAs are dimensioned for an average person with a weight of 75 kg and a shoulder breadth measurement of 400 mm. Our identification of lack of space harmonizes with the research done by (Kozey et al. 2008/2009). Based on measurements of offshore workers in Eastern Canada wearing marine abandonment immersion suits they recommend downgrading the rescue craft capacity by approximately 15% to accommodate the actual size of occupants wearing insulated PPE.

26.3.6   Rescue Craft Moisture

Moisture in the insulation layers of PPE reduces its effective insulation value and has a detrimental effect on the survival rate (Michel B. DuCharme et al. n.d.). All participants were wet when they aborted the exercise. In respect of the participants wearing survival suits, the moisture came from their own body's perspiration. The participants wearing only life jackets experienced moisture accumulating in their clothing from the condensation inside both rescue crafts. This moisture inside the life raft caused great concern as it condensed on the inside of the canopy and accumulated on the floor of the raft where people were sitting.

26.3.7 Additional Stress Factors

Prior to the trial, all participants were briefed on the risks involved and the safety system in place.

When a walrus appeared in the exercise area, only a few meters from the raft, the participants in the life raft had to keep a sharp lookout, and the canopy had to remain open for a prolonged period. Normal routines also had to be abandoned. This diverted the participants' focus from staying warm and resulted in a few participants having to abort the exercise.

On board the lifeboat, one person had to stay outside for some time to assemble the radar reflector, usually a short and uncomplicated task. Due to the cumbersome survival suit, neoprene gloves, cold metal parts and snow on the deck, this job took longer than usual. The participant also had to remove his gloves to complete the task, resulting in cooling of the extremities and degraded fine motor control. Despite returning to the lifeboat, he did not recover the use of his hands and had to abort the exercise some time later.

The ability to manage additional tasks will in many cases cause additional stress. The majority of the participants were focused on staying warm. In a cold climate survival situation, conducting additional tasks that divert the focus from staying warm, will reduce the probability of survival.

26.3.8 Psychological Aspects

In a real situation, the motivation to survive will likely be stronger than in an exercise scenario, but there will also be additional stress factors. All participants expressed the importance of a well-trained lifeboat/life raft captain. This person has a key role in establishing routines and distributing the available resources. The captain of the rescue craft also has an important role in creating routines and predictability. This is of key importance for remaining motivated and utilizing the individual resources in a sustainable manner.

Confident leadership will greatly influence the survival rate of those on a rescue craft. The longer the stay in the craft, the more important is the leadership.

26.3.9 Personal Protection

Assessing the different PPE based on time spent in the LSA gives an indication of the relative functionality of the equipment and how well it protects the participants. See Table 26.1: Personal Protective Equipment for more information. The different types of PPE offered different levels of protection, however, it is clear that the survival suits gave a major advantage over the different types of vests.

Table 26.1 Personal Protective Equipment – the hours that people stayed in the rescue crafts utilizing different personal protective equipment

	Survival Suit Neopren	Survival Suit Insulated	Survival Suit non-Insulated	Thermeal protection West	Kamp-vest with bag	Kamp-vest no bag	Nordkap Drakt	Survival Suit 307
Average Life boat (h)	22.3	22.3	16.0	11.0	15.2	10.0	24.3	
Average Raft (h)	7.6	17.5	14.4	6.4	8.6	6.0	13.2	9.4

While the number of participants within each test condition makes it difficult to state findings as being statistically significant, the results still point to performance gaps between protective survival equipment currently approved through SOLAS and what is now required by the Polar Code.

The large discrepancy for the neoprene survival suit between the lifeboat and the life raft was due to water ingress in the suits. The leaks were not experienced as a problem in the lifeboat, while in the life raft the leaks caused wetness, with a loss of insulating capability in the layers of clothes.

Stochastic studies predict a 50% probability of survival when immersed to the neck in 5°C water for about 3 h in heavy seas, wearing a long-sleeved shirt, light sweater, and jacket (Tikuisis and Keefe 2005). Few studies have however been conducted, investigating the long term effects of heat production caused by shivering response, and there are limited predictive models for long-term exposure to cold (Xu et al. 2005). Significant individual variations with regards to the ability to produce heat induced by the body's shivering response represents a large spread in the data material.

26.4   Conclusions

The trial described here was the first of its kind to be carried-out in the field since publication of the IMO Polar Code. Results suggest that there are gaps in performance for survival equipment currently approved by SOLAS compared to what is required by the Polar Code. It is clear that individual motivation and knowledge play an important role in a survival scenario. Conducting simple tasks like unzipping the survival suits at regular intervals for ventilation and drying out the insulating layer can greatly influence the outcome for that individual. The Polar Code states that equipment is to protect the passengers/crew from hypothermia. When utilizing standard SOLAS approved equipment, compliance with the functional Polar Code requirement of protection from hypothermia is not expected beyond 24 h of exposure in relatively benign polar conditions. With few exceptions, all of the participants had reached the abortion criteria well before 24 h. In a real accident scenario, the participants would have survived for an extended period beyond this point, but for how long is uncertain.

It is very unlikely, however, that a majority of the participants would have survived inside the LSAs for another four days, due to continued loss of core temperature and few opportunities for heat generation. To increase the survival rate, modifications to the functionality of the equipment would be required. These include:

- Higher degree of insulation in the personal protective aids
- A defined level of insulation in survival craft to balance the expected heat loss and ventilation needs for extended survival in polar regions
- Increased space per person to enable movement to ensure blood circulation
- CO_2 measurement devices/alarms inside the rescue craft
- Active ventilation systems to ensure a safe microclimate inside the rescue craft
- Larger and extended range of food and water rations
- Enhanced training of lifeboat/liferaft captains for long term survival situations in polar regions

References

ABS (2016) IMO polar code advisory. Huston, American Bureau of Shipping (ABS)

Andrew B et al (n.d.) Carbon dioxide accumulation within a totally enclosed motor propelled survival craft. Memorial University/National Research Council Institute for Ocean Technology, St. John's

Brunvoll R (2015) MASTERPLAN Svalbard mot 2025. Sysselmannen Svalbard. Hentet fra http://www.sysselmannen.no/Documents/Svalbard_Miljovernfond_dok/Prosjekter/Rapporter/2015/14-%2020%20Masterplan%20Svalbard%20mot%202025.pdf

DuCharme ME (2007) Effect of wetness and floor insulation on the thermal responses during cold exposure in a life raft. In: Proceedings of 12 international conference on Environmental Ergonomics, Piran Slovenia

DuCharme MB, Evely K-A, Basset F, MacKinnon SN, Kuczora A, Boone J, Mak L (n.d.) Effect of wetness and floor insulation on the thermal response during cold exposure in a liferaft. Defence R&D Canada, Quebec City

International Maritime Oranization (2016) POLAR code, international code for ships operating in polar waters. IMO Publishing, London, ISBN 978-92-801-1628-1.

Kozey JW et al (2008/2009) Effects of human anthropometry and personal protective equipment on space requirements. Occupational Ergonomics 8:67–79

Solberg et al. (2016) SARex Spitzbergen, Search and rescue exercise conducted north of Spitzbergen, Exercise Report. University of Stavanger, Stavanger. ISSN 0806-7031/ISBN 978-82-7644-677-7. Hentet fra https://brage.bibsys.no/xmlui/handle/11250/2414815

Tikuisis P, Keefe AA (2005) Stochastic and life raft boarding predictions in the Cold Exposure Survival Model (CESM v3.0). Defence Research and Development Canada, Toronto

Xu X, Tikuisisb P, Gonzaleza R, Giesbrechtc G (2005) Thermoregulatory model for prediction of long-term cold exposure. Computers in Biology and Medicine 35:287–298

Chapter 27
Safety of Industrial Development and Transportation Routes in the Arctic (SITRA) – Collaboration Project for Research and Education of Future High North Experts

Nataliya Marchenko, Rocky Taylor, and Aleksey Marchenko

Abstract Industrial development in the Arctic enhances the potential risk of accidents occurring under severe conditions. Detailed knowledge of the physical environment and understanding of risk reduction methods are necessary for technical experts and young specialists planning to work in companies dealing with the Arctic. The Arctic is a place of close contact between many countries, where harsh and fragile environment demands the most advanced technology for sustainable development and international collaboration to ensure safety of industrial activity. The SITRA (Safety of Industrial Development and Transportation Routes in the Arctic, 2015–2018) project focuses on organizing an international research and educational network of High North experts for joint investigation and teaching of Arctic engineering courses. SITRA is funded by the Norwegian Centre for International cooperation in Education (SIU). The project is a part of the High North Program. It continues the more than 20-year-long Norwegian-Russian collaboration in the field of Ice Engineering and expands it overseas by means of students and staff/professors exchange and joint field work. Canadian and US universities have also joined the team. The SITRA project multiplies the understanding and awareness of the Arctic problems through education and outreach.

N. Marchenko (✉) • A. Marchenko
The University Centre in Svalbard, Longyearbyen, Norway
e-mail: natalym@unis.no; alekseym@unis.no

R. Taylor
Memorial University of Newfoundland, St. John's, NL, Canada
e-mail: rocky.taylor@card-arctic.ca

© The Author(s) 2017
K. Latola, H. Savela (eds.), *The Interconnected Arctic — UArctic Congress 2016*,
Springer Polar Sciences, DOI 10.1007/978-3-319-57532-2_27

Fig. 27.1 SITRA map. Project participants (from west to east): University of Alaska Fairbanks (Fairbansk, US), Dartmouth College (Hannover, US), Memorial University of Newfoundland (St. Johns, Canada), Norwegian University of Science and Technology (Trondheim, Norway), The University Centre in Svalbard (Longyearbyen, Norway), State Marine Technical University of Saint-Petersburg (Russia), Moscow Institute of Physics and Technology (Russia), Lomonosov Moscow State University (Russia)

27.1   SITRA Partners and Objectives

The SITRA project builds up an international team of experts from eight universities of Norway, Canada, US and Russia specialized in Ice Engineering (Fig. 27.1) and providing unique education in the Arctic.

Project activities include teaching and study on courses at the University Centre in Svalbard (UNIS), performance of joint PhD and MSc projects, joint field and laboratory work, numerical modelling, workshops, and international conferences. The teaching methodology follows the research field-based education strategy of the host institution – UNIS.

The main objective of the project is to increase basic knowledge in the following safety aspects: (1) estimation of ice actions on offshore and coastal structures, (2) description of dangerous ice phenomena for navigation in the High North regions, (3) probabilistic methods of risk estimates and practical methods of accident risks reduction. The project aims at organizing joint lecture courses about the physical environment in relation to running industrial activity in the Arctic. Important elements of collaboration are student exchange and joint field work in Svalbard and the Barents Sea.

Project partner's PhD/MSc students visit partner universities for studies and joint research. International cooperation in education increases the flux of knowledge of students significantly. Students get more information about the ongoing industrial projects and research activities focused on safety issues that arise through both industrial activity and navigation within ice-infested waters.

Joint field work helps students to master modern equipment and gain skills necessary for safe work in the Arctic. Collaboration with the large research projects allows to use the expensive devices and attracts experienced researchers.

The project focuses in organizing lectures on:

1. fundamentals of ice mechanics and engineering applications;
2. hydrodynamics of ice-covered waters and its applications;
3. safety problems of offshore and coastal structures in the Arctic;
4. safe navigation in Arctic straits.

All main project activities are reflected on the project website (UNIS and SITRA 2017).

Lectures are organized at UNIS as part of four existing courses, available on the UArctic Study Catalogue (UArctic 2017a): Physical Environmental Loads on Arctic Coastal and Offshore Structures, Ice Mechanics, Loads on Structures and Instrumentation, Arctic Offshore Engineering, and Arctic Offshore Engineering -Fieldwork. Detailed course descriptions are available in UNIS online course catalogue (UNIS 2017). Professors from partner universities visit UNIS as guest lecturers for a period of 1–2 weeks for teaching and field activities.

27.2 Project Implementation and Development in 2015–2016

The project activities in 2015–2016 followed the project goals and plans:

- Three field work series (March, April, November) with full scale ice mechanical tests yearly performing by international research team with participation of UNIS course students;
- MSc student projects;
- Lecturing of professors from partner universities;
- Publication and presentation of the results on the conferences.

International group of scientists extended the experiments that were started in 2013–2014: investigation of mechanical properties of ice (testing of the new rig for ice beam tests, *in-situ* identation tests, tests for compression and tensile strength of ice) and ice actions on coastal structures (coal quay in Kapp Amsterdam), investigation of drag forces on ice and under ice turbulence (ADV measurements, CTD, and ADCP profiling), and investigation of tides in Svalbard fjords. Students of the UNIS course had unique possibility to participate in these scientific experiments, work with state-of-the Art equipment and get experience of Arctic research with well-qualified instructors. See (Murdza et al. 2016; Chistyakov et al. 2016) as an example of the work and results.

Spring investigations and teaching in the field have been continued during the UNIS study cruise and survey on the vessels Bjørkhaug (2015) and Lance (2016), including icebergs observation and towing attempt, tests on fast ice.

MIPT MSc student made comparison of ice strength properties collected from experiments with indentation and uniaxial compression and worked with demo version DE software ITASCA PFC 2D (ITASCA 2017) for the modeling of the towing of a floating structure in broken ice, in the frame of collaboration with the company "Kvaerner Concrete Solutions" project in 2015–2016. In the group of researchers/supervisors he performed the test with L-shaped cantilever beam for complex shear and bending strength (Murdza et al. 2016).

In November 2015, a joint research group, including 2 professors from MSU and 1 from SMTU, led by Arctic Technology department of UNIS made full scale tests of fresh ice on the lake near Longyearbyen. UNIS students took part in the work as a part of the study, they worked in four groups, performing indentation test in the lab and *in-situ*; beam test *in-situ*, fracture toughness test *in-situ*, uniaxial compression test in the lab, analysis of thin sections of ice in the lab; computer modelling of beam test in Comsol Multiphysics (COMSOL 2017). In 2016 due to warm autumn and unfrozen lake, the same experiments were carried out in laboratory conditions. That gave the new experience.

Investigation of wave propagation below the ice registered during the join expedition with C-CARD/MUN Group in 2014 was performed in 2015 (Marchenko et al. 2015a).

Altogether 12 scientific papers (Murdza et al. 2016; Chistyakov et al. 2016; Marchenko et al. 2013, 2015a, b; Mohammadafzali et al. 2016; Marchenko 2015, 2016; Konstantinova et al. 2016; Karulina et al. 2016; Sakharov et al. 2015; Marchenko and Marchenko 2015; Marchenko and Onishchenko 2015; Kowalik et al. 2015) have been published as a result of collaboration, which will be continued in the frame of SITRA project in 2017–2018.

27.3 Beam Test Modelling – Example of a Joint Student Work

Flexural failure of sea ice is of interest in many different shipping and industrial development applications in Arctic regions, ranging from understanding rubble formation processes to modeling bending failure of ice sheets against sloped structures and ship hulls. As part of the study program, students worked together with researchers on investigating mechanical properties of sea ice. Figure 27.2 demonstrates the test with fixed ends beam. In this test, ice beam is prepared by sawing of two ice-through cuts along the beam axis. In the test the beam is broken by the load (indicated by letter F), applied to the beam in the horizontal direction as it is shown on the scheme (left down of Fig. 27.2). The load F is applied to the beam by a hydraulic indenter. The indenter consists of two vertical cylinders (indenter and root cylinder) connected with each other by two horizontal hydraulic cylinders connected to oil pump station. The indenter is mounted on a steel frame with sledges.

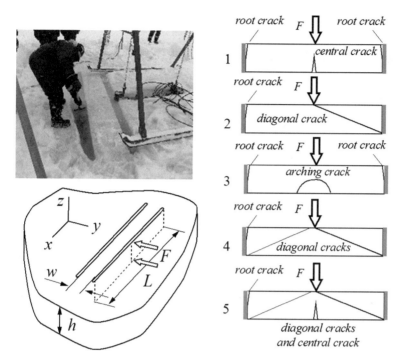

Fig. 27.2 Organizing of the *in-situ* test with fixed ends beam (*top left*) schematic of the test (*right*), and observed ice failure modes (*below left*) (Photograph by Aleksey Marchenko)

Failures and strengths of ice were studied. Five different scenarios of the beam failure shown in Fig. 27.2 from the right were observed depending on the length to thickness ratio of the beam. Bending strength and compressive strength of ice both are calculated from the record of the load in the test if the beam follows the first failure mode. This scenario is well reproduced by finite element modeling. Four other scenarios with formation of diagonal and arching cracks were observed (Mohammadafzali et al. 2016), but not explained and reproduced by the modeling. Results of *in-situ* tests performed in joint expedition by RV Lance in Store Fjord in 2014 were used to parameterize a discrete element model of ice fracture under flexural loading. Simulations of these experiments in 3D were carried out using a new material model within the open-source Discrete Element Method (DEM) code WooDEM (WooDEM 2017) which features cohesive bonds in tension, shear, flexure and torsion based on a contact model with normal, shear, torsional and flexural springs. A comparison of simulated and field test results, such as those in Fig. 27.3 below, along with recommendations for future work were provided (see Mohammadafzali et al. 2016).

Figure 27.3 (Top right) demonstrates ice failure mode 5 reproduced by WooDEM. Two MSU students used a commercial code ITASCA PFC 3D (ITASCA 2017) during their study at UNIS in the autumn 2016. Figure 27.3 (*Bottom left*)

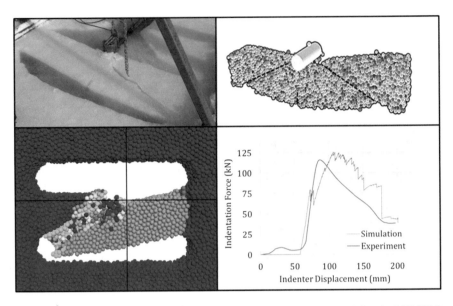

Fig. 27.3 *In-situ* beam test result (*Top left*); Results of discrete element modeling by MUN PhD Student (*Top right*); Simulated DEM results produced by UNIS/MSU MSc students (*Bottom left*; Force-displacement curve comparing simulation and field experiment results (*Bottom Right*)

demonstrates ice failure mode 2 reproduced with PFC3D. Numerical simulations of the test currently continue in close cooperation between UNIS-MSU-MUN.

27.4 Future Development and Perspective

The SITRA project is helping to increase the knowledge needed to ensure Arctic safety throughout the High North. In spring 2017, Canadian students will come to UNIS to participate in Arctic Technology courses and project work, under joint supervision. Canadian and US professors will also teach at UNIS in 2017. The main challenges the project has faced until now have been logistic, connected to remoteness and harsh weather conditions on Svalbard, as well as challenges associated with arranging funding, coordinating the time schedules of students and professors, and resolving visa issues, particularly for the international students. The Arctic Engineering Thematic network in the framework of the University of Arctic (UArctic 2017b) also provides a valuable connection point for this initiative. Even more intensive collaboration with this Thematic Network could prove fruitful for the future project development.

Acknowledgments The authors are grateful to SIU and the colleagues in the project.

References

Chistyakov P, Karulin E, Marchenko A, Sakharov A, Lishman B (2016) The tensile strength of saline and freshwater ice in field tests. In: Proceedings of 23rd IARH international symposium on ice (IAHR-ICE 2016), AnnArbor, Michigan, USA

COMSOL (2017) COMSOL Multiphysics®. The platform for physics-based modeling and simulation. https://www.comsol.com/comsol-multiphysics

ITASCA (2017) PFC – general purpose destinct-element modeling framework. http://www.itas-cacg.com/software/pfc

Karulina M, Marchenko A, Sakharov A, Karulin E, Chistyakov P (2016) Experimental studies of fracture mechanics for various ice types. In: Proceedings of 23rd IARH international symposium on ice (IAHR-ICE 2016), AnnArbor, Michigan, USA

Konstantinova M, Marchenko A, Karulina M, Sakharov A, Karulin E, Chistyakov P (2016) In-situ investigations of ice deformations and loads in indentation tests. In: Proceedings of 23rd IARH international symposium on ice (IAHR-ICE 2016), AnnArbor, Michigan, USA

Kowalik Z, Marchenko A, Brazhnikov D, Marchenko N (2015) Tidal currents in the western Svalbard Fjords. Oceanologia 57(4):318–327

Marchenko NA (2015) Ship traffic in the Svalbard area and safety issues. In: The 23rd international conference on port and ocean engineering under Arctic conditions (POAC 2015), Trondheim, p 11

Marchenko N (2016) Comparison of sea ice products and data of drifting buoys. In: Proceedings of 23rd IARH international symposium on ice (IAHR-ICE 2016), AnnArbor, Michigan, USA

Marchenko NA, Marchenko AV (2015) Sea currents and ice drift in western part of Barents Sea. A comparison of data from floating and fixed on ice buoys. In: The 23rd international conference on port and ocean engineering under Arctic conditions (POAC 2015), Trondheim

Marchenko A, Onishchenko D (2015) Analytical modeling of passive turn of turret moored vessel in close ice. In: The 23rd international conference on port and ocean engineering under Arctic conditions (POAC 2015), Trondheim, p 11

Marchenko N, Zhmur VV, Shkhinek KN, Løset S, Marchenko AV (2013) Safety of Maritime operation and sustainable industrial development in the Arctic (SMIDA). In: 11th international conference and exhibition for oil and gas resources development of the Russian Arctic and CIS continental shelf (RAO/CIS Offshore 2013), Saint Petersburg, Russia

Marchenko AV, Gorbatsky VV, Turnbull ID (2015a) Characteristics of under-ice ocean currents measured during wave propagation events in the Barents Sea. In: The 23rd international conference on port and ocean engineering under Arctic conditions (POAC 2015) Trondheim, p 11

Marchenko AV, Kowalik Z, Brazhnikov D, Marchenko NA, Morozov EG (2015b) Characteristics of sea currents in navigational strait Akselsundet in Spitsbergen. In: The 23rd international conference on port and ocean engineering under Arctic conditions (POAC 2015), Trondheim

Mohammadafzali S, Sarracino R, Taylor R, Stanbridge C, Marchenko A (2016) Investigation and 3D discrete element modeling of fracture of sea ice beams. In: Arctic technology conference

Murdza A, Marchenko A, Sakharov A, Chistyakov P, Karulin E, Karulina M (2016) Test with L-shaped Cantilever beam for complex shear and bending strength. In: Proceedings of 23rd IARH international symposium on Ice (IAHR-ICE 2016), AnnArbor, Michigan, USA

Sakharov A, Karulin E, Marchenko A, Karulina M, Sodhi D, Chistyakov PV (2015) Failure envelope of the brittle strength of ice in the fixed-end beam test (two scenarios). In: The 23rd international conference on port and ocean engineering under Arctic conditions (POAC 2015) Trondheim, p 11

UArctic (2017a) UArctic study catalogue. http://education.uarctic.org/studies/courses/
UArctic (2017b) Thematic networks and institutes. Arctic Engineering. http://www.uarctic.org/organization/thematic-networks/arctic-engineering/
UNIS (2017) What would you like to study? http://www.unis.no/courses/
UNIS, SITRA (2017) Safety of industrial development and transportation routes in the Arctic, 2015–2018 (SITRA). http://www.unis.no/research/arctic-technology/ice-mechanics/sitra-project-page/
WooDEM (2017) Custom Didcrete Element Solutions. https://woodem.eu/

Chapter 28
Safe Snow and Ice Construction to Arctic Conditions

Kai Ryynänen

Abstract Snow and ice are the key elements in winter tourism, especially in the Nordic countries. Finnish Lapland has a multitude of attractions that provide visitors unique arctic experiences in snow and ice constructed environments. In Finland, snow and ice operators and builders are usually small and medium size enterprises (SME). There is lack of knowledge on using snow and ice as construction material. This article will provide basic knowledge on using snow and ice as construction material in safe structures.

28.1 Introduction

Snow and ice are pure natural materials of the Arctic area. They can be used as a building material within certain boundary conditions. Due to its crystal structure, snow – especially artificial snow – is mainly suitable for compressive structures. The density, i.e., volumetric weight of snow in a structure changes during the entire service life of the structure (RIL 2001).

Detailed information on the material properties of snow and utilisation of these properties is available in literature. The material property values that are used in designing snow structures are chosen case by case by the structural engineer. The property values are affected by temperature, snow quality and the load of the snow and of the built structure (RIL 2001). The special properties of snow and ice materials must be taken into consideration in calculations and in construction. The structural safety against breakage, falling over and deformation should be analysed during planning process.

Ice has a special property; its crystal structure does not remain constant under loading or thawing (Ryynänen 2011). The properties of ice change according to the temperature of the ice. The structural properties of ice are weakest at 0 °C. As the

K. Ryynänen (✉)
Lapland University of Applied Sciences, Rovaniemi, Finland
e-mail: kai.ryynanen@lapinamk.fi

© The Author(s) 2017

K. Latola, H. Savela (eds.), *The Interconnected Arctic — UArctic Congress 2016*, Springer Polar Sciences, DOI 10.1007/978-3-319-57532-2_28

temperature of ice rises, its strength decreases and its modulus of elasticity, shear modulus and creep strain grow (Kilpeläinen and Mäkinen 2003).

Slush can also be used as a building material in snow and ice construction. Slush is a mixture of snow and water. Snow and ice structures in which the amount of water added during the construction phase exceeds 5% of the amount of snow used can be called slush structures. The proportion of the different components in the mixture is unevenly distributed (Ryynänen 2011).

28.2 Designing Snow and Ice Structures

The basic principle in designing snow and ice structures is to ensure the safety of the people who are using the structure. The construction guidelines only deal with structures that are large enough or used in a way to require an official permit for their construction. A person who is using calculations done with the formulas presented here must be a professional in the field of construction and hold sufficient competence in structural design (RIL 2001). In Finland, structural design complies with the regulations and guidelines provided in the Finland's National Building Code, issued by the Ministry of the Environment and the Eurocode Standards ratified by Finland (Ministry of Environment 2001, 2002, 2004).

The special properties of snow and ice materials must be taken into consideration in calculations and in construction. The following design parameters must be analysed when designing load-bearing structures:

- Ultimate limit state calculations; these indicate structural safety against breakage and falling over.
- Serviceability limit state calculations; these indicate structural safety against deformation.

The partial coefficient of safety -method is used in the structural calculations. Use of partial coefficients of safety in the calculations produces reliable limit values, so that structural deformations and changes in loading during the service life can be predicted (RIL 2001).

The impact of weather conditions on snow and slush structures must be assessed. In the Finnish construction guidelines, there are instructions how to assess weather conditions. If the construction site and structures are the same every year, weather condition information from previous years, if based on sufficiently reliable measurements, can be used to assess the structural behaviour (RIL 2001).

In serviceability limit state analysis, limit values are set for the structure. If these values are reached or exceeded, the use of the snow and ice structure must be suspended or terminated. Limit values in the serviceability limit state are tilt, sag, window and door functionality, and retention of the original structure (RIL 2001; Ryynänen 2011) as presented in Fig. 28.1.

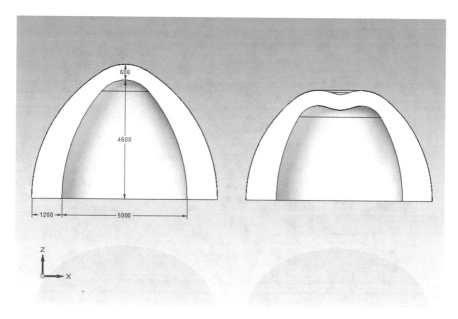

Fig. 28.1 Retention of the original structural shape in snow, ice and slush structures (Ryynänen 2011)

28.3 Designing of Snow Structures

In snow structures, the structural calculations follow the guidelines provided in RIL (2001). When using the guidelines, the density of snow is between 400 and 820 kg/m3. Deformation of snow should be taken in the account. By using snow as construction material, compression structures can be made (Ryynänen 2011). For example, a snow structure designed according to RIL guidelines is shown in Fig. 28.2.

28.4 Designing of Ice Structures

The loads of ice structures are calculated as described in the Snow Structures guidelines RIL 218-2001 (RIL 2001). The density of the ice used in calculations is 920 kg/m^3 regardless of whether natural or artificial ice is used. The thawing effect of the sun and the eroding effect of wind must be taken into consideration in specifying structural thicknesses (Ryynänen 2011). Ice structures are designed at a temperature of 0 °C, because various studies have indicated that the material properties of ice are most reliable and adequate safety during the service life is achieved (for example the tilt and minimum thickness). Ice structures can be designed by using pieces of ice or by freezing (Ryynänen 2011).

Fig. 28.2 Arch snow tunnel (Photograph: Kai Ryynänen 2011)

28.5   Designing of Slush Structures

Loads of the slush structures are calculated as described in Ryynänen (2011). Changes in tensions and deformation which occur during the service life of the slush structure calculated according to the RIL 218-2001 guidelines.

In Finland, construction of buildings or structures intended for professional use – such as in the tourism industry – requires a planning or building permit. This especially applies to cases where structures are constructed from snow or ice and people will be underneath or inside the said structures during their service life. In Finland, the building supervision authority specifies the necessity and type of permit. The objective is always to ensure the safety of the users of the structure (Ryynänen 2011).

The building permit may be granted for a fixed-term or temporary building and for one time or for a five-year period, for example. The designer of a snow and ice building must be a sufficiently competent person with necessary training in the sector and experience in similar design jobs. Finnish authorities who supervise construction and use have specified that the following guidelines must be followed in snow and ice structures (Ministry of Environment 2001, 2002; Ryynänen 2011).

28.6   Safe Use of Snow and Ice Buildings

In case of a fire or other emergency, it must be possible to exit the building safely. For that, the building must have an adequate number of suitably situated, sufficiently wide and easily navigable exits. Structures used as lodging or meeting rooms must be equipped with exit signs. Maps of exit routes must be situated in visible places in the building. In addition, every evacuation area in the building that is occupied by employees or other persons more than temporarily must have at least two separate appropriately situated exits (Pitkänen 2003).

Smoke alarms must be installed to all bedrooms of the snow and ice buildings (Pitkänen 2003) and those fire detectors should be designed for cold climate conditions. If necessary, the building must be equipped with adequate and appropriate fire extinguishing equipment.

Snow and ice buildings are usually places to where everyone must have access. The rooms must be designed so they can be used by disabled persons. Thus, attention must be paid to travel routes, ramps and toilet facilities (Ministry of Environment 2004). Attention must be paid to at least the following items in connection with snow and ice construction: stairs, handrails, guardrails, adequate lighting and anti-slipping measures (RIL 2001).

28.7   Monitoring of the Structures During Their Service Life

Snow and ice structures must be monitored during their service life. Monitoring includes both daily maintenance inspections and long-term monitoring, such as the measurement of deformation. In Finland, a separate operating and safety plan is compiled for each snow and ice building site as a part of its construction (Ryynänen 2011).

Deformation of snow and ice structures must be monitored during the structure's entire service life. The deformation measurement points and measurement intervals for each load-bearing structure must be specified in the building plans. The hardness of the structure must also be monitored during its service life by measuring it at regular intervals. Hardness is monitored with measurements and by visual observations. If one of the limit values (for example whether the structure has sagged or tilted, or fractured/collapsed) is reached during the service life of a snow and ice structure, the use of the structure must be suspended. If the suspension lasts for a long time or the cause cannot be eliminated, dismantling of the structure should be considered (Ryynänen 2011).

The user and the party responsible for maintaining the snow and ice structure have to specify the time when the structure will be dismantled. The authorities may also appoint an external party who specifies the conditions for dismantling of the structure, for example in sites that involve special procedures such as break down roofs and close entrance using snow or ice (Ryynänen 2011).

28.8 Next Steps in Snow and Ice Construction

There is a need for co-operation between the countries such as Finland, Sweden, Norway, Canada, Japan and China that have snow and ice constructions. Ideally, they should build up a network of snow and ice research institutions, universities and snow and ice companies. One aim of such network would be to identify the similarities in the general designing and construction rules between the different countries. Also, there is a need for more research, for example, in the field of ice physics and slush construction, as well as in creation of new types of snow and ice structures.

New learning environments should be developed through innovation activities and co-operation between educational institutions, for example in a form of joint courses in snow and ice construction and snow physics. In the future, attention should also be paid to develop new innovations in the snow and ice construction and to strengthen the opportunities for SMEs, especially in the snow construction and tourism-based businesses.

References

Kilpeläinen M, Mäkinen S (2003) Structural properties of ice. NorTech Oulu report 2/2003. University of Oulu, Oulu

Ministry of the Environment (2001) National building code. F2 building's operational safety. Regulations and guidelines 2001, Helsinki

Ministry of the Environment (2002) National building code. A2 building's designers and designs. Regulations and guidelines 2002, Helsinki

Ministry of the Environment (2004) National building code. F1 accessible building. Regulations and guidelines 2004, Helsinki

Pitkänen M (2003) Building supervision's perspective of snow construction, safety factors. City of Kemi Building Supervision, Kemi

RIL Finnish Association of Civil Engineers (2001) Design and construction guidelines for snow structures. RIL 218-2001. General copy – Painopörssi, Helsinki

Ryynänen K (2011) Snow and ice construction guide, hints for authorities – guidelines for builders. Rovaniemi University of Applied Sciences, report C 27

Chapter 29
The Components of Psychological Safety of Oil and Gas Shift Workers in the Arctic

Yana Korneeva, Tamara Tyulyubaeva, and Natalia Simonova

Abstract The chapter focuses on justification of the psychological safety model of oil and gas workers in the Arctic. The safety in industrial activity depends firstly not only on the employee, on his attitude toward observance of occupational safety and health regulations, but also on the personal attributes of the specialist, his subjective perceptions and effectiveness of his psychological self-regulation. The study was conducted at an oil and gas facility with a watch-based method of labor organization in the Arctic zone of the Russian Federation (duration of a rotation shift is 30 days). The study involved 70 persons at the ages from 24 to 60 years (average age 38.7 ± 1.3). The methods were as follows: study of documentation, monitoring of work process, questionnaires, psycho-physiological and psychological testing, and statistical methods of data analysis. The study verified the concept of psychological safety as a mental state of a subject who has control over a set of internal and external factors of the ergatic system providing updating of internal resources of the individual for efficient professional activity on the psycho-physiological and psychological levels. As a result, the model of psychological security of oil and gas workers in the Arctic was introduced and evidence based. It includes the following components: (1) The psycho-physiological level of functional status (reduced or optimal); (2) The psychological level of functional state (emergency or economical); (3) The image of the labor object (low undifferentiated hazard assessment or high differentiated hazard assessment); (4) The perception of the subject (high undifferentiated or moderately high differentiated self-assessment); (5) The perception of the subject-object and subject-subject relations (neutral, negative, or positive). In this research, the components of psychological safety of oil and gas workers of different professional groups (operators of oil and gas, boiler operators, drivers, engineers and technical workers, maintenance specialists) in the Arctic were empirically studied and characterized. The psychological level of functional state was expressed in economical adaptive strategy mainly on the basis of results gained from the study of operators of treatment facilities and boiler houses, engineering

Y. Korneeva (✉) • T. Tyulyubaeva • N. Simonova
Northern (Arctic) Federal University named after M.V. Lomonosov,
17 Northern Dvina Embankment, Arkhangelsk 163000, Russia
e-mail: ya.korneeva@narfu.ru; musya1991@bk.ru; n.simonova@narfu.ru

© The Author(s) 2017
K. Latola, H. Savela (eds.), *The Interconnected Arctic — UArctic Congress 2016*,
Springer Polar Sciences, DOI 10.1007/978-3-319-57532-2_29

and technical staff and maintenance staff. On the contrary, the emergency strategy was defined in the course of the study of oil and gas operators. Low psycho-physiological levels of functional states were noted among drivers and operators of treatment facilities and boiler houses. Higher rates of this component of psychological safety were found among engineering and technical personnel, oil and gas operators, as well as maintenance specialists. Among the components of psychological safety related to the mental structure of the regulation, there are no strong differences between workers belonging to one group or another, depending on job. In all occupational groups there is a predominance of a higher and more differentiated assessment of the impact of adverse factors and potential hazards during the rotation shift and moderately high self-assessments of competence. However, the composition of the subject-object and subject-subject relations shows a predominantly negative evaluation of socio-psychological environment mostly among drivers, in other groups of specialists the attitude was neutral or positive.

29.1 Introduction

According to the World Health Organization, accidental deaths rank third after cardiovascular diseases and cancer (Sudak 2011). Every year in the world there are 125 million industrial accidents resulting in 1.1 million deaths globally, out of which 25% under the action of harmful and hazardous substances, according to estimates of the International Association of Social Security and the International Labor Organization (Sudak 2011). In the European Union, there are about seven million cases of occupational injuries on a yearly basis (Sudak 2011). The recorded rate of industrial accidents in Russia is 3–12 times lower than in the EU, but the rate of fatal accidents is 3–9 times higher (Sudak 2011). Achieving a high level of industrial safety at all enterprises of extractive industry including oil and gas production is one of the priorities of the "Energy Strategy of Russia for the period until 2020" (Morozov et al. 2013).

The safety in industrial activity depends firstly on the employee. Not only by his attitude toward observance of occupational safety and health regulations, but at the same time on the personal attributes of a specialist, his subjective perception and the effectiveness of his psychological self-regulation. A personal factor in the accident rate is defined as a totality of all mental and physical characteristics of the person, which may be tied to the incident (Kotik & Emelyanov 1993). Therefore, both (1) organization and work environment and (2) characteristics of psychological safety as a potential way to avoid errors should be taken into account by research of employees' safe behavior.

Analysis of previous investigations (Azizov and Khodjaeva 2010; Kora 2012; Korneeva et al. 2013; Kotik 1990; Kozlov 2007; Morozov et al. 2013; Sautkina 2015; Simonova 2011) have revealed a contradiction between scientific evidence of human factors' influence on accidents and emergencies in the oil industry, and small number of studies reflecting psychological characteristics of rotational specialists who committed errors leading to accidents. Another contradiction appears between the need to improve professional activity efficiency of rotational workers, and the lack of systematic analysis of the effectiveness' criteria. One more challenge is an increase of accidents, injuries and occupational errors made by rotational workers, and the lost opportunities to improve the situation by means of psychological follow-up and support.

The psychological factors causing these contradictory findings among these different studies may be adjusted by detailed study of specialists in various professional areas working on a rotational principle of work organization in the Far North.

The objective of this study was to develop a model for psychological safety of oil and gas workers with a rotational principle of work organization in the conditions of group isolation in the Arctic environment.

29.2   Theoretical Justification of the Psychological Safety Model

In this study, the psychological safety implies the mental state of a subject who controls a set of internal and external factors of his/her ergatic system. It provides activation of internal resources of an individual for efficient professional activity on psycho-physiological and psychological levels. Put that way, the meaning of "security" in psychological terms serves as a criterion for psychological safety.

Considering the closeness of both terms to each other, it is necessary to determine how a mental state is understood in psychology.

Mental states are one of the components of the psychological phenomena that include mental processes and mental characteristics (Leontiev 1993).

The activity approach (according to A.N. Leontyev (Leontiev 1993), and S.L. Rubinstein (Rubinstein,2000)) considers mental states as functional, i.e. "The state of a man at work". Functional states are regarded as effective aspects of activity.

Altogether, the subject's internal features are determined by the concept of performance efficiency. A.B. Leonova (Rubinstein 2000) and V.I. Medvedev (Leonova 1984; Rubinstein 2000) consider the efficiency of economic activity at three levels:

1. physiological, which takes into account a state of health, specific features of metabolic processes, as well as neurodynamic properties of the subject;
2. psychological, which considers the functional content of work load, the profession requirements to functional systems which ensure the fulfillment of work tasks (work and rest, work posture, work load);
3. behavioral, which is characterized by the experience, skills, operating style, adaptive strategies of behavior.

M. Greenwood, J. Woods, N.L. Shlykova et al. (Shlykova 2004) suggest the dependence of psychological safety and safe behavior on the professional environmental conditions. Therefore, this study takes into consideration the understanding of mental labor regulators, which could act as components of psychological safety.

E.A. Klimov (2004) defined performance capability as a "functional system" of subject-object relations, and a person (the subject of labor) as the initiator of activity (Klimov 2004).

E.A. Klimov (2004) also identified psychological regulators of labor and psychological criteria that reveal professional development of a worker and his competence's level.

The psychological regulators of labor can be divided into three groups: "image of the object", "image of the subject" and "image of the subject-subject and subject-object relations".

All labor regulators refer to the "image of the subject" group, which is formed during professional activities by gaining experience.

"The image of the object" helps to evaluate the worker's perceptions of labor process as a whole and its structural components (subject, objectives, tools, working environment, duties, etc.).

"The image of the subject" is an actual "self-image" (self-identity) revealing the self-consciousness level of a worker, i.e. self-perception, awareness of one's features, interests, values and meanings in accordance with the professional reality where he/she is acting; awareness of his/her role in the professional community and in the society.

"Image of subject-object and subject-subject relations" is an indicator used to explore and assess the level of professional self-consciousness, which formation is regulated by needs, emotions and feelings, attitude toward different perspectives of the objective reality, personality orientation and world outlook.

Given that psychological safety is understood as mental state which is considered to be a functional state according to activity approach, functional state both on psycho-physiological and psychological level can act as components of psychological safety by constructing a psychological safety model. In addition, the psychological safety model can include psychological labor regulators: "image of the object," "image of the subject" and "image of the subject-subject and subject-object relations" are seen through the regulation psychological structure (Table 29.1).

Table 29.1 Psychological safety model

Psychological safety	
Functional status (Leonova 1984; Rubinstein 2000)	**Mental regulation of activities** (E.A. Klimov 2004)
1. Psycho-physiological level	1. The image of the labor object
2. Psychological level	2. The image of the labor subject
	3. The image of the subject-object and subject-subject relations

29.3   Materials and Methods

The study was conducted at the oil and gas facility with a 30-day rotational shifts organization in the Nenets Autonomous District of the Russian Federation. The study involved 70 persons in the ages from 24 to 60 (average age 38.7 ± 1.3 years). The length of work experience by rotation method in shifts varied from 0.5 year to 31 years (average experience 9.53 ± 1.2 years).

The methods used in the study were as follows: study of documentation, monitoring of work process, questionnaires, psycho-physiological and psychological testing, and statistical methods for data analysis. The questionnaire was designed to obtain information on the life record data of employees and the specific features of their work. The questionnaire included the following sections:

- general information on the education and work experience;
- subjective assessment of the adverse climatic and geographical, industrial and social factors that affect the workers during rotational shift;
- specific organization of free time during rotational shift;
- subjective assessment of the individual's professional effectiveness and level of professionalism;
- subjective assessment of different dangerous situations that may arise during the rotational shift;
- subjective assessment of workplace hazards and factors that contribute to their formation.

To achieve the research objectives, the following methods were used:

1. Complex visual-motor response (CVMR) using a psycho-physiological testing device UPFT-1/30 "psychophysiology" (Ltd. Medicom MTD, Russia). This is an assessment of the operator's performance by parameters of dual-alternative complex visual-motor reactions.
2. Variable cardio measurement (VCM) using a psycho-physiological testing device UPFT-1/30 "psychophysiology". This is an assessment of the functional condition and adaptive capacity of the cardiovascular system according to the method of cardio measurement.
3. Measurement methods of cerebral hemispheres activation using hardware and software complex "Aktivatsiometr AC-6" (INGOs "Acceptor", Russia).
4. A questionnaire "The state of health. Activity. Mood"(Karelin 2007) developed by V.A. Doskino, A.N. Lavrent'ev, V.B. Sharay and M.P. Miroshnikov (Karelin 2007). The questionnaire is intended for assessment of the state of health, activity and mood.
5. Color preferences test (M. Lüscher adaptation Sobchik L.N.) (Sobchik 2001). The test is aimed at identifying emotional and characterological basis of personality and its current state thereof. In order to use the test data of M. Lusher, interpretational factors developed by G.A. Amineva were applied.
6. "Express-method" for investigation of working social-psychological environment by O.S. Mikhalyuk and A.Y. Shalyto (Dmitriev et al. 2010). This method

Table 29.2 Methods and techniques of diagnostics components of psychological safety of rotational workers

Components of psychological safety	Methods and techniques
Psycho-physiological levels of functional states	1. A complex visual-motor response, using a psycho-physiological testing device UPFT-1/30 "psychophysiology" (Ltd. Medicom MTD, Russia)
	2. Variable cardio measurement, using a psycho-physiological testing device UPFT-1/30 "psychophysiology" (Ltd. Medicom MTD, Russia)
	3. Measurement of activation, using hardware and software complex "Aktivatsiometr AC-6" (INGOs "Acceptor", Russia)
Psychological level of functional states	1. Questionnaire "State of health. Activity. Mood" (Sobchik 2001)
	2. Color preferences test (Dmitriev et al. 2010)
Image of labor object	1. Study of documentation
	2. Observation of workflow
	3. Survey of environmental factors and dangerous situations arising during the rotational shift
Image of labor subject	Questionnaire about the self-assessment as a professional.
Image of subject-object and subject-subject relations	"Express-method" to study the socio-psychological environment in the workplace (Dmitriev et al. 2010)

facilitates identification of emotional, behavioral and cognitive components of relations in the work team during the rotational shift.

Correlation between the methods, the diagnostic techniques and the components of psychological safety of rotational workers is presented in Table 29.2.

The statistical methods used in results' analysis were as follows: descriptive statistics; contingency tables with calculation of Pearson X2 criterion; two-step cluster analysis. Statistical analysis was performed using the statistical package IBM SPSS Statistics (license agreement № Z125-3301-14 (Northern (Arctic) Federal University).

29.4 Results and Discussion

Based on the theoretical analysis, functional status and psychological labor regulators compound psychological safety of a person. To study the psychological safety components, a two-step cluster analysis was carried out for the variables related to the psycho-physiological and psychological parameters of functional states, the image of the object and the subject of labor, as well as for the image of the subject-object and subject-subject relations. The results show two clusters corresponding to professional groups having a statistically significant difference per totality of psychological security parameters (Table 29.3).

Table 29.3 Components of the psychological safety of oil and gas workers identified as a result of the empirical study

The psychological level of the functional status		
Emergency	Economical	
Psycho-physiological level of the functional status		
Reduced	Optimal	
The image of the labor object		
Low and undifferentiated risk assessment	High and differentiated risk assessment	
The image of the labor subject		
High and undifferentiated assessment	Moderately high and differentiated assessment	
The image of the subject-object and subject-subject relations		
Neutral	Negative	Positive

Based on the results, one can assume that the reliability of professional activities is related to the development of psychological safety. It can be traced by high levels of physiological parameters of a worker, a positive attitude to the working team on the emotional, cognitive and behavioral levels. Not less important are choice of adaptation strategies, presence of a moderately high self-appraisal of one's professional skills and the adequate assessment of adverse factors and potential hazards encountered by shift workers in the Far North.

Acknowledgements   The study was sponsored by the Russian President's grant for state support of young Russian scientists – PhD (MK-7500.2016.6).

References

Azizov HF, Khodjaeva GK (2010) Risk analysis of accidents oilfield pipe systems Nizhnevartovsk district. Bull Nizhnevartovsk State Univ 1:50–53

Dmitriev MG, Belov V, Parfenov Yu (2010) Psycho-pedagogical diagnosis of delinquent behavior among troubled teens. Saint Petersburg

Karelin AA (2007) Big encyclopedia of psychological tests. Moscow

Klimov EA (2004) Introduction to the psychology of work. Moscow

Kora NA (2012) The psychological aspect of the category "safety". Kazan Pedagog J 3(93):67–72

Korneeva YA, Simonova NN, Degteva GN, Dubinina NI (2013) The adaptation strategy of shift workers in the Far North. Hum Ecol 9:9–12

Kotik MA (1990) Psychology security activities, from the first publications in Dorpat to current research in the University of Tartu. Knowledge and regulation of activities: historical, developmental and applied problems: Ouch. Rec Tartous Univ 894

Kotik MA, Emelyanov AM (1993) The nature of the human operator error: on transport control example, Moscow

Kozlov MM (2007) Development and improvement of methods of increasing safety of oil and gas industry workers on the basis of the method of registration of dangerous situations, Moscow

Leonova AB (1984) Psychodiagnostics of human functional states. Moscow

Leontiev AN (1993) About the Book N. Rubinstein "Fundamentals of general psychology" (comments AA Leontiev, DA Leontiev and MG Yaroshevsky). Psychol Mag 4:21–29

Morozov IS, Krivetsky IM, Volokhina AT, Glebova EV (2013) Analysis of the features of professional work of staff of LLC "Gazprom mining Nadym", affecting the safety of the production process. In: Proceedings of the Russian State University of oil and gas. THEM Gubkin 4 (273), 132-142.

Rubinstein SL (2000) Fundamentals of general psychology. St. Petersburg

Sautkina EA (2015) Analysis of occupational injuries in the OJSC "Tyumenenergo". In: Proceedings of the international participation of the scientific and practical conference of students, graduate students and young scientists.

Shlykova NL (2004) Psychological safety of the subject of professional activity. Tver

Simonova NN (2011) Psychological analysis of professional activity of specialists of oil and gas complex (on example of shift labor in the Far North). Moscow

Sobchik LN (2001) Method of color choices. A modification eight-color M. Luscher test. Practical guide. Saint Petersburg

Sudak SN (2011) Analysis of occupational injuries in Russia and the Murmansk region for the period 2005-2009. Bull Murmansk State Tech Univ 4:860–866

Part VI
Circumpolar, Inclusive and Reciprocal Arctic

Chapter 30
Finding Gender in the Arctic: A Call to Intersectionality and Diverse Methods

Gunhild Hoogensen Gjørv

Abstract The following chapter examines multiple aspects of including gender perspectives in Arctic research. In the chapter I discuss the definition and understanding of the concept of gender, and then move to the concept of "intersectionality" which recognizes the important linkages between multiple identities of gender, race, ethnicity, class, age, and other social categories. I then discuss both the ways in which gender has been addressed, though still minimally, in Arctic research, as well as some of the ways in which Arctic research is itself gendered. I then discuss how gendered perspectives add important insights into understanding security, and more specifically human security, in the Arctic.

At the UArctic conference in St Petersburg in September 2016 I was struck, yet again, by the distinct absence of gender data and analyses, even though this conference was yet another arena whereby Arctic scientists could come together and share their continued insights into Arctic societies and environments. Though it would be unfair to claim that absolutely no one included gender insights and analysis into their research presentations at that event, there was no question that gender was not central or key to many or most research projects. Anecdotal observations and discussions with conference participants were revealing, as, when I asked some presenters about gender aspects to their research (in this case education and psychology respectively), they acknowledged the relevance of gender to their own work but did not consider raising gender as an issue in their research and presentations. Their reaction and approach was not uncommon as I continued to inquire amongst colleagues. The experience prompts me to ask, "where is gender in the Arctic?" And does our broad Arctic research community have an adequate understanding of what it means to include gender perspectives in Arctic research?

In this chapter I will address a number of issues surrounding the implementation of gender perspectives in Arctic research. I will first focus on what we mean by "gender" and further "intersectionality", that informs my argument. I follow with a discussion about the gendered nature of Arctic research, and how gender has been

G. Hoogensen Gjørv (✉)
UiT The Arctic University of Norway, Tromsø, Norway
e-mail: gunhild.hoogensen.gjorv@uit.no

© The Author(s) 2017
K. Latola, H. Savela (eds.), *The Interconnected Arctic — UArctic Congress 2016*,
Springer Polar Sciences, DOI 10.1007/978-3-319-57532-2_30

situated in Arctic research since the Taking Wing conference in 2002. Lastly I will use my own research to briefly illustrate how an intersectional approach (including gender, race, ethnicity, class, age, etc) is crucial to both understanding and implementing multidisciplinary and cross-cutting Arctic research.

30.1 Definitions of Gender and Intersectionality

"Current Arctic discourses reflect a very masculine Arctic agenda" (Retter 2015)

Integrating "gender awareness" or gender perspectives in the context of Arctic communities and research has been increasingly in focus since the early 2000s (Health 2002). Despite more than a decade of explicit attention, I would argue that gender perspectives are nevertheless still poorly understood as both a category and method in research, and thus generally marginalized in Arctic research. In this section I will draw upon gender and feminist studies literature from the field of international relations as well as upon broader feminist and gender studies research.

Gender is a primary social category used to define social and political relations. It is used in all societies though often in different ways. In general however, is rooted in, but goes beyond, the biological perception of the two sexes, male and female. The concept refers to socially constructed identities and differences between men and women, reflected in characteristics assigned to categories of "masculine" and "feminine". These roles are context and time specific, whereby different cultures and histories have their own interpretation of "man" and "woman"; the characteristics of masculine and feminine can be performed by persons of any sex category (male, female, and others), thus critiques of either masculinity or femininity are not critiques of "men" or "women" per se, but of the values, norms and practices these categories have been constructed to embody. The relationships between the resulting categories of masculinity and femininity are in constant negotiation and renegotiation (Skjelsbæk and Smith 2001).

Gender is thus a central social and political dimension by and through which human societies are based. Even those, whose research is focused on environmental changes and developments within marine, terrestrial, atmospheric or cryospheric aspects of the Arctic, have either direct or indirect impacts on human societies, including transfers of toxins into human food chains, climate change resulting in potential harms to humans such as floods, avalanches, melting of permafrost, changes in marine and terrestrial animal migration patterns, changes in ecosystems, etc. These changes can and do impact people differently based on their gender and gender roles, depending on the society in question. Societies that reflect gendered values and practices also impact Arctic science.

While Arctic science has been dominated by research methods and practices that reflect masculinist values of rationality and objectivity (often reflected in positivism), gender and feminist research has been moving beyond these research parameters to provide more comprehensive analyses of the complex relationships between the social and natural worlds. One important move in this direction was the development of the concept "intersectionality" recognizing that universalizing, homogenous methods and practices were often both inacurrate as well as harmful to research as well as to the societies that were central to such research. As well, universalizing definitions of gender equality and understanding of gender constructions across all societies were grossly inadequate. The three waves of feminism were dominated by experiences of generally white, middle-class, Euro/Western women, and these experiences did not speak to either the gendered norms, practices or experiences of people of colour, indigenous people, non-white-centric ethnicities and cultures, nor to those with differing experiences based on age, class, sexuality, and ability (Marfelt 2016). Coined by Kimberlé Crenshaw in the late 1980s (Crenshaw 1991), the term "intersectionality" was designed to critically assess the intersection between race and gender, and at its core has a "non-positivistic, non-essentialist understanding of differences among people as produced in on-going, context-specific social processes" (Marfelt 2016: 32). For the rest of this chapter I will refer to the broader, more methodologically inclusive approach of intersectional analyses unless I am referring specifically to gender.

30.2 Intersectionality and Science

Understanding the ways in which Arctic science has been conducted is crucial to understanding how and to what degree intersectional analyses have been integrated into this body of scholarship. Intersectional research is generally part of the broader domain of social sciences, which itself has experienced "science wars", a contestation of methodologies, methods, and approaches (Keating and Della Porta 2010). Social sciences operate at high levels of abstraction, where social inquiry includes exploring ontologies, epistemologies, approaches, methodologies, and methods without predetermining the process of inquiry (ibid). As such, different processes of inquiry result in different constructions and productions of knowledge.

Feminists have long demonstrated a gendered and masculinist bias within concepts and approaches to scholarship, not least exemplified by the emphasis on rationality, objectivity, and public domains, often embodied by research in the natural sciences and visibly expressed in un-reflexive, silent authorship reinforcing "an unreflective orientation toward objectivist traditions and norms" (Gray 2017: 180). A core feature of feminist and intersectional methodological approaches therefore includes the practice of "reflexivity" whereby the researcher is "'responsible' and 'responsive' to her work and her 'subjects' of study because it makes explicit the deliberative movement of her scholarship" (Ackerly et al. 2006: 258, cited in Agathangelou and Turcotte 2008). Reflexivity allows for insight into phenomena while also illuminating how such insights were derived: "the closer an academic discipline is aligned with the natural science model the greater the pressure can be to engage in un-reflexive silent authorship" (Gray: 182). Thus, the dominance of a natural science heavy Arctic scholarship informed by objectivist methods plays a significant role in the acceptance and comprehension of what intersectional analyses bring to the discussion. My point is not to discredit objectivist/positivist types of study, as these bring necessary knowledge to light. However, Arctic research would benefit from a more substantial engagement with a plurality of methods and voices. Disciplines using an intersectional, reflexive approach have been largely relegated to "niche" research areas few are expected to tangle with, despite calls for "cross-cutting" research between social and natural sciences. However, Arctic research is strengthened by providing complex insights into the broader social and political contexts in which all Arctic research takes place.

30.3   A Glimpse into Arctic Intersectional Research

Intersectionality tells us that gender, race, ethnicity, class, etc. are central social and political dimensions by and through which human societies are based. As noted above, research focused on environmental changes and developments within marine, terrestrial or cryospheric aspects of the Arctic, have either direct or indirect impacts on human societies. Even though there is already evidence that environmental, social, and political change can and does impact men and women differently, does the marginalization or absence of gender-focused resesarch in Arctic societies, in all their diversity, mean that Arctic societies have gender issues "figured out"? Or that Arctic policy is always "beyond" gender and it therefore does not need attention?

The still minimal, but important, intersectional research done on Arctic issues and communities would indicate that the answer to those questions should be a resounding "no". Indeed, insofar as we see gender-inclusive research, it is still quite focused on gender itself, rather than engaging an explicitly intersectional approach. In 2002, the Arctic Council conference on Gender Equality and Women in the Arctic "Taking Wing" took place, addressing a range of issues from women in the workplace including the heavily male-dominated extractive industries, living conditions, traditional knowledge and self-determination for indigenous peoples and the impacts on indigenous women, political participation, health, and violence against women (Health 2002). As the first gender-focused conference hosted by the Arctic Council, the organizers already recognized the importance of gender as a cross-cutting issue, set against the backdrop of climate change and its impacts on ecosystems, globalization, cultures and peace and justice. In other words, gender in the Arctic was clearly about taking an intersectional approach, even before the notion of intersectional began to take a foothold in feminist analytical literature.

There have been a number of events or initiatives focusing on gender since 2002, including the two Arctic Human Development Reports (Einarsson et al. 2004; Larsen and Fondahl 2014), and the Gender Equality conference in 2014 which took place in Akureyri, Iceland (Oddsdóttir et al. 2015). There has been a special issue on gender published by *Anthropology of East Europe Review* in 2010, but like many of these other initiatives, and as I mentioned above, one can quickly get the impression that gender is a special interest or niche area rather than a broader source of data and an analytical resource and tool. Arctic focused conferences have generally had less of an emphasis on gender, let alone intersectionality. Ideally I, as a researcher focused in part on gender, as well as others, could argue that the reason for a lack of explicit focus on gender or intersectionality is because researchers are integrating gender analyses as a cross-cutting theme. This does not seem to be the overall case however. A quick review of more recent Arctic-themed conferences like Arctic Science Summit Week (ASSW), Arctic Frontiers, Arctic Circle, and even the International Congress of Arctic Social Sciences (ICASS) has thus far had limited focus on gender or other intersectional issues. Indeed, amongst 22 themes which consist of roughly 5–15 sessions each, originally only one session at the 2017 ICASS would explicitly addresses gender in broader understandings of Arctic societies. At the recent International Arctic Science Committee (IASC) meetings in Prague, CZ, a new gender working group was initiated, which has worked to ensure that gender will be better represented at ICASS 2017 in Umeå in June 2017, with both a panel session and a roundtable debate about gender research and gendered methods in Arctic science. Hopefully this initiative will have a spillover effect to future Arctic conferences. A review of the previous 2016 program of Arctic Circle shows that gender was not once mentioned in the program, and an email to the Arctic Frontiers secretariat asking about this was never answered.

As a further example of integrating gender perspectives into Arctic research, we can look to the first Arctic Human Development Report (Einarsson et al. 2004). Though rife with more questions than details (Oddsdóttir et al. 2015) the report was a groundbreaking achievement as an excerise in intersectionality, combining a diverse and wide-ranging group of Arctic researchers together towards creating a comprehensive overview and understanding of natural, social, and political life in the Arctic. Though not explicitly highlighting intersectionality *per se*, the ADHR chapter clearly illustrated the tensions between different understandings of gender in the Arctic and that one approach does not fit all. That chapter resulted in having the most co-authors of any other chapter, not least because of the broad range of issue areas that are relevant, but also because the impacts of gendered social constructions and their resulting impacts on political, social and environmental life are not universal across Arctic geographies and Arctic communities. There could not have been a better beacon to expand gender analysis in the Arctic, given the complex dynamics the ADHR Chap. 11 illustrated.

The second AHDR (Larsen/Fondahl 2014) made an explicit decision to demand that all chapter contributors address gender as a cross-cutting issue in each of the different chapter topics that were raised in the report. Not all chapters were equally successful. Remarkably the Economic Systems chapter (Chap. 4) made no mention of gender and mentioned women as a socio-economic category only briefly in one textbox. The Resource Governance chapter equally made no mention of gender, mentioning women only once, in reference to forest owners (Larsen and Fondahl 2014: 281). Indeed, an additional quick check for keywords "men", "transgender", "two-spirit" and "LGBTQ" achieved, unfortunately, no results. As noted in the 2002 "Taking Wings" report, as well as the first AHDR, economics and resources have played a significant role in the development of gender roles in communities in the

Arctic. Natural resource and extractive industries are heavily male dominated to this day (World Bank 2015), and these industries have also been argued to be based upon masculinist value systems of exploitation that often have negative effects on the environment and climate (Miller 2004; Kawarazuka et al. 2017). The Arctic provides a hotbed of cases where extractive industries do, and will continue to contribute, to negative environmental impacts, and understanding the basis of these systems is crucial to any movement towards change.

Surprisingly, the chapter on cultures and identities was remarkably devoid of gender analysis or discussions about the role of gender in culture and identity aside from stating that more research was needed (Larsen/Fondahl 2014: 143, 144). The chapters on populations and human health and well-being were quite strong in addressing gender issues. However in general, an overriding comment amongst many of the 2014 ADHR chapter authors was that research on gender analysis was lacking and that future research needs to include these analyses (ibid). Much of the discussion revolved around statistical data as well, and did not delve deeper into the values embedded in different Arctic societies that are linked to economic development strategies, marginalization of peoples and decolonization practices, and relationships of people to the environment. This evidence does not present a strong case that we have come much farther than the 2002 "Taking Wing" conference.

30.4   Moving Towards Intersectionality in Arctic Security Research

How can we use an intersectional approach to illuminate social, political, and environmental developments taking place in the Arctic? In my own work on the Arctic I use both a intersectional and a security analysis approach. Both concepts are actively present politically and socially across the region, and they are intimately tied to environmental use (or exploitation) and change. The term security invokes power, whereby the utmost priorities of the person, state, social group are linked to the survival of values and practices for the future (Hoogensen Gjørv 2017). Many are quite familiar with the narrower, militarized understanding of security that focuses on the use of the military for purposes of defending one actor, the state, from existential threat (Walt 1991). However, when we think in terms of actors more broadly, and diverse practices to ensure survivial (and the diverse ways in which survival is understood), a much more complex analysis results. Through intersectional analysis, it is possible to best understand the dynamics and tensions between priorities and perceived futures for the Arctic.

This can be exemplified with the concept of "human security" which has been increasingly employed in Arctic research to articulate linkages between individual and community values and prerequisites for survival, and political and social policy. Human security was popularised in the 1994 United Nations Human Development Report, expanding the notion of security to include dimensions of food, health, community, environmental, economic, personal and political security, with the intention to, in part, address some of the glaring weaknesses of security theory and practice. Human security focuses on the individual as its referent. Though actively employed and debated for over two decades, there is no consensus upon a definition. Nevertheless, human security as a concept has had staying power, and is now being used in relation to the Arctic, not least demonstrated by the first edited volume that links environmental and human security in the Arctic to better understand the dynamics between nature and human, and between humans, through gendered analyses (Hoogensen Gjørv et al. 2014).

Environmental and human security (which is often informed by intersectional analyses) are concepts that continue to defy a fixed definition and have been, and continue to be, subject to controversy. Should environmental issues be "securitized"? Defining environmental security can engage an intersectional approach, as the definition of environmental security encapsulates a competition for power about whose priorities and values are heard – the one who is able to define environmental security can further inform practices and policies, including such policies as the COP Paris Agreement of 2015. Which definition of environmental security is employed can mean the difference between the extent to which fossil fuel industries continue to extract oil and gas reserves to the detriment to environments and socio-political communities, or if such activities can be controlled based on a broader understanding of security that includes indigenous and gender-based (eco-feminism) priorities (Ingólfsdóttir 2016).

In what ways is the environment related to humans and human security, if at all, and in what ways are such connections perceived as legitimate (and by whom)? Security is, even amongst many critical security studies scholars, a concept that embodies an understanding of immediate threats that requires urgent action, and such immediacy is often not clear and present when it comes to the environment (Buzan et al. 1998). Due to the history of security concept, environmental security has often been "militarized" and masculinized through the addition of conflict scenarios arising from environmental catastrophes or degradation (Homer-Dixon 1991, 1994), though the thesis has been heavily critiqued (Gleditsch 1998). Many scholars acknowledge that "securing" the status quo of today's western lifestyles is largely contrary to the goals of "securing" the environment, while sustainable development, often characterized as ecological security, suggests that current modernization practices are not sustainable, and therefore a possible threat to security (Krause/Williams 2003). Economic development and the environment, both important to security, are pitted against one another, and little in the way of a solution appears in sight.

In the Arctic, economic security cannot be completely isolated, for example, from environmental security or political security. Food security is connected to environmental, health, and economic security, and so on. Even if we restrict ourselves to environmental security and never mind human or energy security, for example by concerning ourselves only with the environmental (atmospheric and oceanic) linkages between the polar climates and the rest of the planet, these too are extremely complex and non-linear in their relationship (SCAR 2005). They also give an incomplete picture without the complexity of the social and political dimensions brought in by human beings.

Examining environmental security in relation to human and ecological security (which prioritizes environmental protection and preservation) moves environmental security away from its masculinist and culturally western state-based roots (defending a destructive modern way of life, propagating environmentally destructive activities through traditional security mechanisms such as the military, etc) and reflects both the mounting concerns about environmental degradation, as well as emphasizes the importance of human relationships to the environment, prioritizing an intersectional focus. Both approaches are necessary to a more effective understanding of environmental security – ecological security stresses the interconnectedness of all elements within an ecological system, both how they impact as well as are impacted by the system. A "widened" environmental security approach brings the ecological and the human together: "… reformulate environmental security in terms of human security and peace, and drawing on the insights of ecological security" (Barnett 2001: 122). Humans are not only threatened by environmental threats, but cause them as well. A gender, and better yet intersectional, analysis, can illustrate this as a struggle between masculinized values of exploitation and feminized values of protection. As emphasized in the 2004 AHDR gender chapter, the ways in which humans interact and understand the systems in which they exist can differ substantially. Thus we have competing views of security.

An intersectional analytical approach has the ability to transcend and integrate many of the levels and sectors of security that scholars have otherwise chosen to analyze separately. Instead of playing into the dominant approaches to security studies which focus on a very small portion of the security grid and from the top down, gender analysis takes its starting point from the bottom up; it reaches all the way down to the individual, as gender analysis acknowledges that even the personal is political, and reflexive where the researcher is also a part of the system, and therefore the individual's experience is relevant. At the same time it is recognized that individuals are part of communities, and that gender is a significant feature of individual identity in relation to others and is therefore a part of societal security (Hoogensen and Rottem 2004). The social constructions of gender come in to play in the analysis, and the ways in which humans have constructed their societies on the basis of gender roles, who has the "right" to play which roles in the society, and how people are supposed to relate to one another. Intersectional analysis has demonstrated not only the dominance of male or patriarch-based societies, but culturally dominant societies, where the gendered demands (for example, Western feminists) of one society are imposed upon other, less dominant societies. Such processes have and continue to take place in the Arctic.

Intersectional approaches have a logical place in the human security discussion, bringing the political "down" to the level of the individual, to bring a voice to the personal. The personal is political, and human security, with its focus on the individual, has the potential to support these personal voices. Discourses and practices are made visible – by looking within, through, behind (closed doors) and beyond those in power - be they the state or powerful research practices and institutions, multiple actors come into view, including those who are often marginalized. Thus intersectionality in the Arctic, which also includes non-western approaches, highlights features of the security dynamic which have been isolated, ignored, and made invisible because the realities of gender and "other" have not been acknowledged. Just as gender is not reduced to "women's issues" in the Arctic, women in the Arctic are not reduced to a unified, monolithic whole, and men and women of the Arctic experience different forms of in/security on the basis of a combination of complex factors. Many of these factors are related to the environment, where livelihoods ranging from industrial to traditional rely upon as well as impact the natural environment. Indigenous women in the North West Territories, for example, may experience insecurities related to impacts of settler colonial relations in combination with economic insecurities generated by lack of opportunities in local economies for men in their families, which in turn exacerbates rates of domestic violence as well as suicides (Irlbacher-Fox 2015). Questions regarding priorities and values about "why it should be preferable to engage in short-term destructive mining activities at the cost of long-term sustainable economies already existing in the North" (Retter 2015) require analysis about the values within the cultures that compete for power in communities – both settler and indigenous. These values are in part expressed through systems that privilege those who adopt a masculinist, and racist, system of knowledge and power. To be able to dismantle this power, we need to understand it, and understand our role in it as researchers.

30.5 Concluding Remarks

This chapter has provided a brief overview of what it means to move from gender awareness to intersectional analysis, as well as a small snapshot of some of the security issues in the Arctic, moving away from traditional security perspectives. Security is examined from the margins or from positions of non-dominance through an intersectional approach. The Arctic is rife with examples of both marginalized regions (although each of the eight countries of the Arctic treat such regions quite differently), with an important focus upon the fact that the entire Arctic is a region inhabited by those who have traditionally been placed at the margins – the indigenous peoples. The experiences in the Arctic are varied, and the (in)securities in one part of the region are not necessarily the same in another part. However there are enough similarities to argue for attention to be drawn to this region and examine how people live and cope in a region with significant and special challenges. The Arctic demonstrates the importance of the environment and the human relationship with the environment to security, particularly human and societal securities with regard to traditional versus market economies, culture and identities based on relations with environment from fisheries communities to indigenous communities, health and food securities – which by their complexity demand intersectional analyses to be better understood.

References

Ackerly BA, Stern M, True J (2006) Feminist methodologies for international relations. In: Ackerly BA, Stern M, True J (eds) Feminist methodologies for international relations. Cambridge University Press, Cambridge, pp 1–15

Agathangelou AM, Turcotte H (2008) Feminist methodologies for international relations. Politics Gend 4(1):184–187

Barnett J (2001) The meaning of environmental security: ecological politics and policy in the new security era. Zed Books, London

Buzan B, Wæver O, de Wilde J (1998) Security: a new framework for analysis. In: Boulder. Lynne Rienner Publishers, London

Crenshaw K (1991) Mapping the margins: intersectionality, identity politics, and violence against women of color. Stanford Law Rev 43(6):1241–1299

Einarsson N, Larsen JN, Nilsson A, Young OR (eds) (2004) Arctic human development report (AHDR). Stefansson Arctic Institute, Akureyri

Gleditsch NP (1998) Armed conflict and the envi-ronment: a critique of the literature. J Peace Res 35(3):381–400

Gray GC (2017) Academic voice in scholarly writing. Qual Rep 22(1):179–196

Health, M. o. S. A. a (2002) In: Tohka L (ed) Taking wing conference report: conference on gender equality and women in the Arctic. Ministry of Social Affairs and Health, Helsinki

Homer-Dixon TF (1991) On the threshold: environmental changes as causes of acute conflict. Int Secur 16(2):76–116

Homer-Dixon TF (1994) Environmental scarcities and violent conflict: evidence from cases. Int Secur 19(1):5–40

Hoogensen Gjørv G, Bazely DR, Goloviznina M, Tanentzap AJ (eds) (2014) Environmental and human security in the Arctic. Routledge, London

Hoogensen Gjørv G (2017) Tensions between environmental, economic, and energy security in the Arctic. In: Fondahl G, Wilson G (eds) Northern sustainabilities: understanding and addressing change in a circumpolar world. Springer International Publishing, Cham

Hoogensen G, Rottem SV (2004) Gender identity and the subject of security. Security Dialogue 35(2):155–171

Ingólfsdóttir A (2016) Climate change and security in the Arctic: analysis of norms and values shaping climate policy. PhD

Irlbacher-Fox S (2015) Political participation of women in the northwest territories (NWT), Canada. In: Oddsdóttir E, Sigurdsson AM, Svandal S (eds) Gender equality in the Arctic: current realities, future challenges conference report. Ministry for Foreign Affairs, Reykjavik

Kawarazuka N, Locke C, McDougall C, Kantor P, Morgan M (2017) Bringing analysis of gender and social-ecological resilience together in small-scale fisheries research: challenges and opportunities. Ambio 46(2):201–213

Krause K, Williams MC (2003) Critical security studies. Routledge, London

Keating M, Della Porta D (2010) In defence of pluralism in the social sciences. Eur Political Sci 9:S111–S120

Larsen JN, Fondahl G (eds) (2014) Arctic human development report: regional processes and global linkages. Norden, Copenhagen

Marfelt MM (2016) Grounded intersectionality: key tensions, a methodological framework, and implications for diversity research. Equality, Divers Inclusion: An Int J 35(1):31–47

Miller GE (2004) Frontier masculinity in the oil industry: the experience of women engineers. Gend Work Organ 11(1):47–73

Oddsdóttir E, Sigurdsson AM, Svandal S (eds) (2015) Gender equality in the Arctic: current realities, future challenges conference report. Ministry for Foreign Affairs, Reykjavik

Retter G-B (2015) Sustainability and development in the Arctic. In: Oddsdóttir E, Sigurdsson AM, Svandal S (eds) Gender equality in the Arctic: current realities, future challenges conference report. Ministry for Foreign Affairs, Reykjavik

SCAR (2005) SCAR Report: No 24, International Council for Science, Editor. Scientific Committee on Antarctic Research, Scott Polar Research Institute: Cambridge

Skjelsbæk I, Smith D (eds) (2001) Gender, peace and conflict. Sage, London

Walt SM (1991) The renaissance of security studies. Int Stud Q 35(2):211–239

World Bank Group (2015) Gender and the extractive industries: an overview, in energy and extractives. World Bank: https://olc.worldbank.org/sites/default/files/WB_Nairobi_Notes_1_RD3_0.pdf

Chapter 31
Towards an Arctic Awakening: Neocolonalism, Sustainable Development, Emancipatory Research, Collective Action, and Arctic Regional Policymaking

Ulunnguaq Markussen

Abstract The Arctic is increasingly subject to processes of global change, presenting new challenges to Arctic peoples. As the world becomes more aware of the importance of the Arctic, the concept of Arctic risk is becoming globalised, advancing technocratic discourses and solutions that suit metropolitan (rather than Arctic) interests. Arctic peoples require knowledge, research, and resources from outside the region yet must take care to avoid economic, educational, and political neocolonialism operating under the disguise of sustainable development. The changing Arctic, however, offers Arctic peoples new opportunities for collective action. We recommend a form of Arctic regional policymaking that works across multiple channels and levels of formality to foster genuine sustainable development that meets the needs of all Arctic peoples. Such collective action should reach across and beyond state borders, bringing together Indigenous peoples and other Arctic communities, as well as cultivating awareness of shared interests through emancipatory research and education.

The Arctic is changing, whether we like it or not.

Changes in the region's environmental, technological, cultural, and political conditions pose new challenges but also new opportunities. These are rooted in part how we – as the peoples of the Arctic – respond to change. We who have the greatest stake in the region must likewise take the lead in this new era, help form and formulate Arctic futures that serve the needs of our communities, cultures, environments, economies, and wellbeing. The peoples of the Arctic are awakening to the positive potential for change. In this chapter, 'peoples of the Arctic' and 'Arctic peoples' refers to both Indigenous and non-Indigenous communities residing in the Arctic

U. Markussen (✉)
Ilisimatusarfik/University of Greenland, Nuuk, Greenland
e-mail: ulum@uni.gl

© The Author(s) 2017
K. Latola, H. Savela (eds.), *The Interconnected Arctic — UArctic Congress 2016*,
Springer Polar Sciences, DOI 10.1007/978-3-319-57532-2_31

region. This diverges from the state-centric perspective (such as that operating within the Arctic Council), which privileges the interests of 'Arctic states' even though the dominant political actors in most such states are not themselves based in the Arctic.

Climate change means melting sea ice, thawing permafrost and, more generally, changing local and regional conditions. These changes will produce both positive and negative effects in the Arctic, from increased navigability of arctic waters to decreased ability to engage in traditional livelihoods. Similarly, the spread of information technology is offering Arctic peoples the possibility not only to learn about developments in the wider world but also to reach out directly to allies across the region and farther afield, thereby bypassing existing communicational intermediaries in the metropolitan centre. By the same token, however, this increased openness and cultural porosity may enhance the power of detrimental demonstration effects, leading to culture loss and greater assimilation into metropolitan society. In addition, the very political structures through which societies have managed change over the past four centuries are themselves undergoing considerable change, as the power of the sovereign state is challenged by growing local, regional, supranational, and global policymaking on the one hand and governance incursions by transnational corporate actors on the other. Some Arctic peoples risk losing the shelter and support of the metropole[1] At the same time, the Arctic peoples have the potential to embrace new opportunities for collective political action and policy innovation.

31.1   Against Neocolonialism

We have become increasingly attuned to the necessity of asserting our rights when dealing with transnational corporations and powerful state actors, and such actors have in turn become increasingly aware of the necessity of taking our rights into account (Wilson 2016; Papillon and Rodon 2017). Foreign capital and expertise is not in itself bad, but it is vital to ensure that foreign-led and financed economic activities in the Arctic serve the needs of Arctic peoples as well in order to prevent becoming caught up in neocolonial processes. From mining companies and fishing interests to government authorities based in distant seats of power, we must seek to simultaneously be open to the world and avoid becoming a passive cog in the great machine of global capital and *realpolitik*.

The globalisation of corporate power is accompanied by a globalisation of risk and of concern for Indigenous peoples and the future of the Arctic. As the world awakens to the dangers of climate change, increasing attention is being paid to scientific research into the Arctic, research that – though often well-meaning – is also often embedded in metropolitan mindsets, which ultimately reinforce existing

---

[1] The metropolitan power; the power centre within the centre-periphery relationships at work in a given polity, such as Denmark relative to Greenland, Moscow relative to the Russian Arctic, or indeed the institutions and processes controlled by such centres of power.

power structures, priorities, and dependencies. Each new scientific report produced from the metropole highlights new challenges for Arctic peoples while advancing ever-more technocratic discourses and globalist solutions. People the world over are at last becoming aware of how important the Arctic is to them, yet this growing recognition is not guaranteed to result in greater influence for Arctic peoples; it might just as easily result in renewed metropolitan efforts to control developments in the region to suit state or globalist interests – perhaps at the expense of those who make the Arctic their home. Indeed, international focus on climate change may in some cases be politically convenient for powerful actors rather than ideally suited for improving the lives of those in the affected regions (Kelman et al. 2015; Baldacchino and Kelman 2014).

We require knowledge, research, and resources from outside the region if we are to undertake sustainable development and take advantage of the changing conditions. However, care must be taken to avoid buying into or submitting to a neocolonialisation of the Arctic. Images of melting glaciers and starving polar bears may galvanise the wider public and power holders to action in a manner that is not beneficial to Arctic peoples.

The peoples of the Arctic are not responsible for having produced global environmental threats, yet metropolitan society calls upon us to forego economic development and extractive industries, take symbolic action against global challenges, and stand at the vanguard of a new era of 'conspicuous sustainability' (Grydehøj and Kelman 2016). Such efforts to persuade us to turn our lands and seas into nature preserves, to transform our traditional territories into a form of World Heritage and present the Arctic as a treasure belonging equally to the world as a whole, are ultimately requests that we who have been so long exploited and so thoroughly dispossessed further sacrifice ourselves for the continued wellbeing of those who have proven unwilling to make sacrifices. Effective efforts to preserve Arctic environments and revitalise Arctic livelihoods should take precedence over symbolic action and 'conspicuous sustainability' that provides little real benefit (Grydehøj and Kelman 2016). We do not ask to live from the charity of metropolitan benefactors. We do not wish to be the realisation of metropolitan visions of happy, hardy Northerners in need of patronage from the forces of global capital. The last thing we want is to be a canary in the coalmine, a cautionary tale (Farbotko 2010).

31.2   A New Era of Arctic Cooperation

If we are to guide change in the Arctic rather than merely become victims of it or take on roles designed to appeal to metropolitan sensibilities, it is necessary to accept that change is occurring and then commit to managing it. The Arctic is home to many peoples, who possess diverse sets of needs. Yet we are united by a number of shared – if not universal – interests, which can be supported through collective action, whether through formal bodies that can engage with sovereign states or through looser, more flexible coalitions of actors. What is clear is that Arctic sustainability depends on long-term governance across national borders transcending single state, corporate, or even ethnic interests, in light of the remarkable interdependence of Arctic peoples and environments surrounding their shared sea.

As Arctic peoples, we must identify what we really want so that we can begin setting common targets, planning for and implementing strategic actions, and taking on strategic responsibilities. By creating a true Arctic community, one that works at multiple levels of formality and through multiple channels, we can at last truly gain a voice (Plaut 2012). An openness to informal political organisation must complement a dedication to constructing strong institutional foundations for Arctic policymaking that reflects shared values and priorities. As we work to define our own ethnic and national identities, we should seek to locate these within a wider Arctic identity, one that can enrich – rather than overshadow – our localised values. Such a multifaceted, pan-Arctic policymaking platform could strengthen our ability to engage with non-Arctic actors (including metropolitan interests within states that possess Arctic territory), which are both affecting and affected by changes taking place within the Arctic. This contrasts with current regional bodies and institutions, which often take a broad state-centric perspective (for example, the Arctic Council and the Barents Euro-Arctic Council) or a narrow focus on Indigenous interests (for example, the Inuit Circumpolar Council and the Indigenous Peoples' Secretariat): Both such perspectives are necessary and important, but the former places agency in the hands of the metropole while the latter risks creating false divisions between Arctic communities, erroneously grouping the interests of Arctic non-Indigenous peoples with those of the metropole.

Arctic economic, societal, cultural, and political development must be led by Arctic peoples themselves, working to develop peaceful and constructive relationships at the community, state, and other levels. A sense of a shared stake in the Arctic can even help foster better relationships between Indigenous peoples and former colonisers, supporting a decolonial Arctic politics that regards all Arctic peoples as equal – understanding the consequences of historic injustice without perpetuating ethnic and political animosity. Understanding today's Arctic and envisioning a future Arctic requires acknowledgment that, across much of the North, "the institutional and ideological interference of the colonial period directly or indirectly, subtly or overtly, continues to influence political thought and economic agendas" (Nadarajah and Grydehøj 2016: 442; see also Grydehøj 2016).

These must not, however, be decisions made by a local elite. It is important that the entire population of the Arctic be involved in the move toward sustainable regional development and come to understand the benefits to collective action and collective development. At the same time, desired development outcomes must be localised; no longer can it be acceptable for Arctic actors to pursue competitive, zero-sum games against one another. At the same time, as we recognise that shared development may require certain shared sacrifices, we must draw the line at sacrificing some communities solely in favour of others.

31.3 Arctic Research and Education

This kind of regional awareness and bottom-up policymaking requires enhanced knowledge on the part of Arctic peoples and an improved connection between science and policymaking. Education is key to this, with actors such as UArctic working to build local capacity through education. Yet it is important to be conscious of neocolonial forces within education systems and to prevent the development of an Arctic cultural elite that is distanced from knowledge of ordinary communities across the region. Development planning is increasingly driven by systemitisation, data collection, and technocratic discourses, which often combine to privilege metropolitan interests and expertise, leading to local suspicion of development and planning in general (Pugh 2013, 2016).

Without denying the importance of 'evidence-based policymaking', it is necessary to recognise that, in today's world, not all kinds of evidence are treated equally. In particular, the experiences of Arctic Indigenous peoples, who have centuries or millennia of experience living in close interaction with nature, can contribute lessons in flexible livelihoods and adaptability to harsh and changing environments (e.g. Pearce et al. 2015). Climate change, for example, will have serious and possibly even disastrous repercussions in some parts of the world, but this change *will* need to be dealt with, and its effects on Arctic environments and peoples may not be straightforwardly negative. It is vital that Arctic policy breaks free from the crippling strictures of neoliberal risk, vulnerability, and resilience discourse (Evans and Reid 2014). Scientific research and evidence-based policymaking that is sensitive to Arctic conditions may reach quite different conclusions than research and policymaking with a metropolitan orientation. The scientifically and technocratically motivated drive toward 'modernisation' asks Arctic Indigenous peoples to submit to a developmental model that has proved environmentally calamitous on the global scale and that has been applied elsewhere with such poor results in earlier periods of decolonisation. Involvement of Arctic peoples in Arctic research is a necessary but insufficient condition for truly locally oriented research: The framework, expectations, and design of the research must itself be decolonised.

Arctic peoples require an emancipatory and decolonial system of research and education, which neither turns its back on knowledge and cutting-edge scholarship nor find itself trapped in preconceived notions regarding the creation of colonial (or postcolonial) subjects. Scientific research itself – from the humanities, social sciences, and natural sciences – should be undertaken by or in close engagement with people from a diverse range of communities, who can both contribute their own perspectives to the wider scientific world and better communicate findings to their own communities (Mulligan and Nadarajah 2008). Greater investment in Arctic science is critical for achieving Arctic sustainability – but only if this science has firm roots in Arctic communities.

Arctic peoples share a number of cultural values but also possess cultural distinctions from one another, which produces a rich seam of traditional and Indigenous knowledge that can be drawn upon in developing Arctic research and education. Truly Arctic research and education does not involve rejecting science or Western learning; it instead involves the investment of such knowledge with local values in order to better serve local needs. To paraphrase Grant McCall (1994), Arctic studies must consider Arctic peoples on their own terms, not on those of metropolitan well-wishers, development workers, or political authorities.

31.4   Towards an Arctic Awakening

Global change is producing new challenges for peoples of the Arctic, but it is also bringing us closer together. We cannot simply draw the line at the individual community, ethnic group, nation, or even state: It is vital for us all to get engaged with the shared Arctic project. By grasping the necessity of change, we can enter a new era of regional stability, shared purpose, and political power.

Change is coming. We may find it daunting, yet it is up to us, the peoples of the Arctic, to say: We will no longer be passive observers of our development but will take charge of our futures.

We belong to the Arctic, and the Arctic belongs to us.

Acknowledgements I would like to thank Adam Grydehøj, my respectful mentor, for his valuable support and contribution on this chapter.

References

Baldacchino G, Kelman I (2014) Critiquing the pursuit of island sustainability: blue and green, with hardly a colour in between. Shima 8(1):1–21

Evans B, Reid J (2014) Resilient life: the art of living dangerously. Polity, Cambridge/Malden

Farbotko C (2010) Wishful sinking: disappearing islands, climate refugees and cosmopolitan experimentation. Asia Pac Viewpoint 51(1):47–60

Grydehøj A (2016) Navigating the binaries of island independence and dependence in Greenland: decolonisation, political culture, and strategic services. Polit Geogr 55:102–112

Grydehøj A, Kelman I (2016) The eco-island trap: climate change mitigation and conspicuous sustainability. Area

Kelman I, Gaillard JC, Mercer J (2015) Climate change's role in disaster risk reduction's future: beyond vulnerability and resilience. Int J Disast Risk Sci 6(1):21–27

McCall G (1994) Nissology: a proposal for consideration. J Pac Soc 17(2–3):1–8

Mulligan M, Nadarajah Y (2008) Working on the sustainability of local communities with a 'community-engaged' research methodology. Local Environ 13(2):81–94

Nadarajah Y, Grydehøj A (2016) Island studies as a decolonial project. Island Stud J 11(2):437–446

Papillon M, Rodon T (2017) Proponent-Indigenous agreements and the implementation of the right to free, prior, and informed consent in Canada. Environ Impact Assess Rev 62:216–224

Pearce T, Ford J, Willox AC, Smit B (2015) Inuit ecological knowledge (TEK), subsistence hunting and adaptation to climate change in the Canadian Arctic. Arctic 68(2):233–245

Plaut S (2012) 'Cooperation is the story'–best practices of transnational indigenous activism in the North. Int J Human Rights 16(1):193–215

Pugh J (2013) Speaking without voice: participatory planning, acknowledgment, and latent subjectivity in Barbados. Ann Assoc Am Geogr 103(5):1266–1281

Pugh J (2016) Postcolonial development, (non)sovereignty and affect: living on in the wake of Caribbean political independence. Antipode

Wilson E (2016) What is the social licence to operate? local perceptions of oil and gas projects in Russia's Komi Republic and Sakhalin Island. Extract Indust Soc 3(1):73–81

Printed in the United States
By Bookmasters